RFID and Wireless Sensors using Ultra-Wideband Technology

Remote Identification Beyond RFID Set

coordinated by
Etienne Perret

RFID and Wireless Sensors using Ultra-Wideband Technology

Angel Ramos
Antonio Lazaro
David Girbau
Ramon Villarino

ELSEVIER

First published 2016 in Great Britain and the United States by ISTE Press Ltd and Elsevier Ltd

ISTE Press Ltd
27-37 St George's Road
London SW19 4EU
UK

www.iste.co.uk

Elsevier Ltd
The Boulevard, Langford Lane
Kidlington, Oxford, OX5 1GB
UK

www.elsevier.com

Notices

Knowledge and best practice in this field are constantly changing. As new research and experience broaden our understanding, changes in research methods, professional practices, or medical treatment may become necessary.

Practitioners and researchers must always rely on their own experience and knowledge in evaluating and using any information, methods, compounds, or experiments described herein. In using such information or methods they should be mindful of their own safety and the safety of others, including parties for whom they have a professional responsibility.

To the fullest extent of the law, neither the Publisher nor the authors, contributors, or editors, assume any liability for any injury and/or damage to persons or property as a matter of products liability, negligence or otherwise, or from any use or operation of any methods, products, instructions, or ideas contained in the material herein.

For information on all our publications visit our website at http://store.elsevier.com/

British Library Cataloguing-in-Publication Data
A CIP record for this book is available from the British Library
Library of Congress Cataloging in Publication Data
A catalog record for this book is available from the Library of Congress
ISBN 978-1-78548-098-0

Printed and bound in the UK and US

Contents

Preface

Wireless sensor networks (WSNs) for smart cities, smart homes and Internet of Things (IoT) applications require low-power, low-cost and simple radio interfaces for a very large number of scattered sensors. UWB in time domain is used here as an enabling radio communications technology.

A comprehensive circuit model is explained for time-coded UWB RFID. Reader setups based on commercial impulse radars are combined with signal processing techniques. As a starting point, several chipless time-coded RFID tag designs are shown as examples. Then, the tags' performance is shown in terms of a number of possible IDs, maximum reading distance, polarization, influence of attached materials, angular behavior and bending (for tags on flexible substrates).

Chipless wireless sensors are derived based on these tag designs. Specifically, for temperature and concrete composition (the latter enabled by permittivity sensing) chipless sensors are shown as possible options.

In order to have more complex, more robust, and longer read range solutions, two chip-based semi-passive sensing platforms are inferred from the chipless tag designs. A wake-up link is used to save energy when the sensor is not being read.

Using an analog semi-passive UWB platform, a wireless temperature sensor (powered by solar energy) and a wireless nitrogen dioxide sensor (enabled with carbon nanotubes and powered by a small battery) are explained. Using a digital (with a low-power microcontroller) semi-passive UWB platform, a multi-sensor tag capable of measuring temperature, humidity, pressure and acceleration is analyzed.

In order to even further increase the read range, two active time-coded RFID systems are illustrated, based on the use of signal amplifiers within the tag.

Finally, a smart floor application for indoor localization is introduced, by joining the proposed designs with ground penetrating radar technology.

Angel RAMOS
Antonio LAZARO
David GIRBAU,
Ramon VILLARINO
February 2016

Acknowledgements

The authors would like to acknowledge the Spanish Government Projects TEC2008-06758-C02-02 and TEC2011-28357-C02-01, the Universitat Rovirai Virgili grant 2011BRDI-06-08, the AGAUR Grant FI-DGR 2012 and the H2020 Grant Agreement 645771– EMERGENT. The authors would also like to acknowledge the undergraduate students Sergi Rima, Xavier Domenech, Eduard Ibars and Cristian Hernandez.

Introduction to RFID
and Chipless RFID

Automatic identification (ID) of goods is widely used in industry, logistics, medicine and other fields. The aim is to obtain the ID information of a good in transit. Giant electronic commerce platforms such as Amazon, Alibaba or eBay are becoming the main choice for buyers worldwide [LOE 14]. Instead of buying from a small retailer, final customers are directly in contact with a wholesaler or distributor. In this context, accurate tracking of each good to its final customer is a major concern in a massive and growing logistics market. An efficient, automatic organization of the stock in large warehouses (both sellers' and logistics companies') is also crucial to reduce costs and delivery times.

Nowadays, the barcode (see Figure 1.1) is the most used automatic ID solution [PAL 07]. It consists of a reader that optically reads a tag. The tag is created by printing black stripes on a white background. Depending on the number, width and separation of stripes, a unique ID is generated. In order to code more information in a smaller space, variations such as QR codes [DEN 14] have recently arisen. The cost of each barcode tag is extremely cheap because it only requires paper and ink. In addition, barcode readers are cheap, and even low-cost compact mobile phone cameras can provide high-resolution images to read barcodes [OHB 04]. However, it requires a direct line of sight between the reader and the tag. A specific positioning of the object is required in order to orientate the barcode toward the reader, and normally only one tag can be read at a time. Barcode storage capacity is also limited, and they cannot be reprogrammed. Another common problem with barcodes is misreading due to a low-resolution printing of the tag, or ink wearing away in harsh environments.

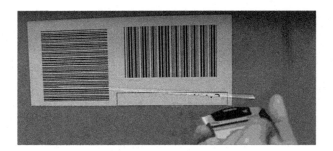

Figure 1.1. *Photograph of a barcode system*

1.1. RFID: state of the art

1.1.1. *Introduction to RFID*

In order to overcome barcode limitations, radio frequency identification (RFID) technologies have been developed in recent years [FIN 10]. RFID systems are used to remotely retrieve data from target objects (tags) without the need for physical contact or line of sight by using magnetic or electromagnetic (EM) waves. With some RFID systems, it is also possible to measure several tags at the same time and rewrite the tag information.

Figure 1.2 shows a basic scheme of an RFID system. There are two main families: near-field RFID (Figure 1.2(a)) and far-field RFID (Figure 1.2(b)) [WAN 06]. Near-field RFID is based on Faraday's principle of magnetic induction (magnetic coupling). Both the reader and the tag have coils. The reader powers up the tag's transponder chip, which can be rewritten. Near-field RFID based on this inductive communication is used for small distances, typically below $\lambda/(2\pi)$ where λ is the wavelength [WAN 06]. ISO 15693 and 14443 standards set frequencies below 14 MHz, which results in a range of a few centimeters. Near-field RFID is widely used for cards and access control, but not for goods management due to its limited range. Far-field RFID uses EM waves propagated through antennas both in the reader and the tag. A reader can be monostatic if it only has an antenna that acts for transmission (Tx) and reception (Rx). On the contrary, if the reader has separate Tx and Rx antennas, it is bistatic. The reader sends an EM wave that is captured by the tag's antenna at a distance of several meters. There are several standards for far-field RFID, with the Electronic Product Code (EPC) Gen2 standard, at the Ultra High Frequency (UHF) (868 MHz in Europe or 915 MHz in the United States) band, being the most used.

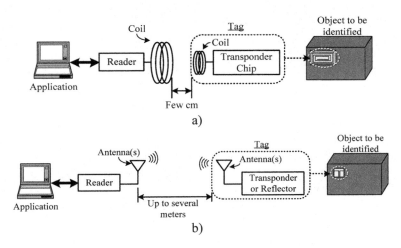

Figure 1.2. *Scheme of an RFID system; a) near-field and b) far-field*

Even though the barcode is still the *de facto* standard, RFID is one of the fastest growing sectors of the radio technology. As of 2014, nearly every commercially available smartphone integrates near-field RFID with the Near Field Communication (NFC) forum's standards [HAR 14]. Wal-Mart and Tesco, some of the largest retailers in the United States the and the United Kingdom, respectively, are adopting RFID [WAN 06]. Furthermore, wireless ID has developed into an interdisciplinary field. Radio frequency (RF) technology, semiconductor technology, data protection and cryptography, telecommunications and related areas come together to develop cheap, secure, reliable, long-range and self-powered RFID tags.

Far-field RFID systems can be classified depending on how the tags get the necessary energy to respond to the readers. Active tags are the most expensive tags, since they need their own power supply (i.e. batteries) not only to power their own chip but also to generate the radio signal with the response to the reader. Semi-passive tags are less expensive than active tags, since they need batteries, but only to power their own logic circuitry, not a transmitter. The response is achieved by changing the reflected signal from the reader in a process called backscattering. This means that the batteries can be smaller and have longer life times (usually years). Finally, passive tags are the cheapest ones and have the largest commercial potential for large-scale spreading [VIT 05, COL 04]. Passive tags use the reader's RF signal to harvest the necessary power for themselves [VIT 05]. Specifically, passive UHF EPC tags are the type of RFID tags most widely used for large-scale applications. Depending on the region, there are different frequency bands and maximum allowed powers allocated for RFID applications [GS1 14]. In Europe, the most used band is 865.6–867.6 MHz, with a maximum transmitted power of 2 W of effective radiated power (ERP), or, equivalently, 3.28 W of effective isotropic radiated power (EIRP). Similarly, in the United States the allowed RFID band is 902–928 MHz, with a maximum transmitted power of 4 W EIRP, or, equivalently, 2.44 W of ERP. It can be observed that American regulations permit more transmitted power than European regulations, allowing for longer read ranges. Most manufacturers provide UHF RFID tags and readers compatible with both European and American bands. Figure 1.3 shows an example of a typical commercial UHF EPC Gen2 reader and tag from Alien Technology [ALI 16]. These types of tags have a sensitivity of about −20 dBm [IMP 14, ALI 14], and read ranges between 6 and 10 m depending on the region [EXT 10]. Recent research has increased the read range to about 25 m by assisting the tag with a battery (battery-assisted passive tags) [ZHE 14].

There have also been recent developments in millimeter wave bands. Millimeter wave identification (MMID) has been presented in [PUR 08] as a concept of RFID operating at 60 GHz. MMID is not a replacement of RFID, since its read range is much shorter (a few centimeters). MMID, however, permits high data rate communications (even gigabit). Directive antennas at millimeter wave frequencies are also very small compared to UHF, permitting the possibility of selecting a tag by pointing toward it. The use of nonlinear devices for RFID tags has also been studied recently. Tags based on the inter-modulation distortion of devices have been presented in [CAR 07] using a diode for localization applications, and in [VII 09] using the micro electromechanical system.

a)   b)

Figure 1.3. *a) Alien ALR-9900 UHF EPC Gen2 RFID reader;
b) Alien ALN9740 UHF RFID tag*

1.1.2. *Chipless RFID*

Chipless tags are a specific type of passive RFID tags. In these tags, instead of storing the ID in a digital IC, it is stored in physical permanent modifications when the tag is fabricated. These modifications change from one tag to another. A notable reduction in costs for passive UHF tags has been achieved recently [VIT 05] due to the popularization in using RFID technology. However, each UHF tag price is fixed by the chip and by the process of connecting it to the tag antenna. Consequently, using chip-based tags is non-viable for identifying large volumes of paper or plastic documents such as banknotes, postage stamps, tickets or envelopes, since the price of the tag is larger than the document itself [COL 04]. UHF RFID technology also presents some weaknesses. UHF frequency-band allocation depends on the region, as well as the readers' output signal power, which directly affects the read range (the more power allowed, the longer the read distance). UHF tags are affected by multipath propagation [LAZ 09a], interference between readers [LAZ 09b] and frequency detuning due to different materials used as the tag physical support [LOR 11], factors that can lead to smaller read ranges. It is also necessary to consider special tags, used when attached to metal surfaces, which increase the total price.

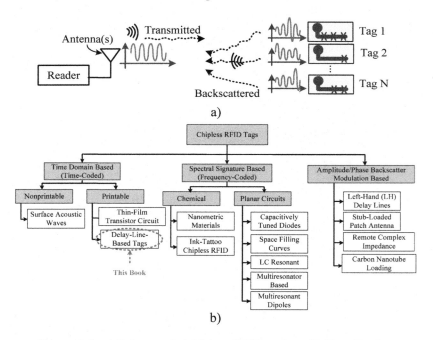

Figure 1.4. *a) Scheme of a chipless RFID system. b) Classification of chipless RFID tags [KAR 10]. For a color version of the figure, see www.iste.co.uk/ramos/rfid.zip*

Chipless tags can be a promising low-cost alternative for RFID systems, since they do not need an IC to work [KAR 10, TED 13]. In chipless tags, the ID is stored in physical permanent modifications in a scattering antenna. The modifications are unique for each tag, and change its RF backscattered response, or signature. Figure 1.4(a) shows a scheme of a chipless RFID system. It is important to note that chipless tags cannot change their information once they have been fabricated, since their physical characteristics are permanent. However, chipless RFID can provide a low-cost alternative, which could increase the capabilities of barcodes. Since a standard for chipless RFID does not exist, there are several types of approaches undergoing active research to achieve chipless RFID tags. Figure 1.4(b) shows a classification of chipless RFID tags given in [KAR 10]. One drawback with chipless RFID tags compared with chip-based tags is the small number of possible IDs [KAR 10, TED 13]. However, this drawback is not very important if the chipless tag integrates additional capabilities beyond ID such as sensing.

Time-domain based (time-coded) tags encode the ID in the time delay of a reflected peak. Surface acoustic wave (SAW) technology offers a nonprintable alternative for chipless RFID [HAR 02, REI 98, REI 01]. SAW RFID is usually based on passive RFID systems, where the signal from the reader is converted into an acoustic wave. A scheme of a SAW tag is shown in Figure 1.5. The acoustic wave hits the tag substrate. Then, multiple reflections occurring at different time instants modify the wave. Then, it is reconverted to an RF signal and sent to the reader. These types of systems have great immunity to temperature changes, have high data transfer rates, can integrate sensors, and have a high read range [HAR 02]. However, SAW tags are expensive and cannot be made easily due to their piezoelectric nature. Therefore, SAW chipless RFID cannot be used with low-cost products [COL 04]. Thin film transistor circuit (TFTC) tags can be printed at high speeds on low-cost films [DAS 06]. They are small in size and have low power consumption. However, manufacturing TFTCs is not a low-cost process, and they are limited to several megahertz. Delay line-based chipless tags consist of an antenna followed by a delay line. Similar to SAW tags, delay line-based tags code the ID in reflections introduced by the delay line. Delay line-based tags can operate either at a narrowband [SHR 07] or wideband [DOU 10] frequencies, but their ID capacity (number of bits for a given tag size) is small. However, the read range of these types of tags is larger than frequency-coded or amplitude/phase backscatter modulation, as will be detailed next. Delay line-based chipless tags at wideband frequencies are studied in detail in this book.

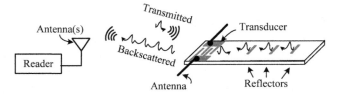

Figure 1.5. *Scheme of a SAW tag. For a color version*
of the figure, see www.iste.co.uk/ramos/rfid.zip

Spectral signature–based (frequency-coded) chipless tags encode the ID using resonant structures. Each bit state corresponds to a presence or absence of a resonance at a given frequency. Frequency-coded tags are printable, robust, have a moderate ID capacity and are low cost. However, a large spectrum is often required in order to encode a large number of IDs, which may not be under regulation at all frequencies. Frequency-coded tags are more sensitive to orientation and distance than delay line based tags, and their read range is shorter. Chemical tags are designed from a deposition of resonating fibers or special electronic ink [COL 06]. In [COL 06], tags fabricated from particles of chemicals that resonate at specific frequencies when illuminated by EM waves are presented. Ink-tattoo chipless tags use electronic ink patterns printed on the surface of the object being tagged: no actual substrate is required [JON 07]. Planar circuit frequency-coded chipless RFID has been the subject of research by several groups. In [JAL 05], a tag consisting of an array of vertically polarized identical dipoles, capacitive tuned, is presented. Each dipole is tuned at different frequencies to code a data bit. In [MCV 06], a frequency-coded tag based on space-filling curves at 900 MHz is presented. Space-filling curves can create resonances with very small footprints compared to the frequencies they are resonating. The main drawback with these types of tags is the difficulty in creating the appropriate layouts to achieve the required resonant frequencies. LC resonant chipless tags consist of a magnetic resonant coil at a particular frequency. Instead of working at a predetermined frequency, as with NFC standards, the reader sweeps a frequency band searching for a resonant frequency peak, which corresponds to the unique tag frequency (ID). Commercial LC resonant chipless tags are widely used for surveillance portals and antitheft purposes at supermarkets and retail stores [FLE 02]. One interesting type of multi resonator-based frequency-coded tag consists of a structure with two antennas in cross-polarization [KAR 10]. The antennas are connected with a transmission line, loaded by resonators at different frequencies. The backscattered response codes the information in the presence or absence of the resonant peaks, determined by the resonators loading the transmission line in the tag. Finally, in [TED 13], another type of multi-resonant structure is presented. In this case, the structure is created by several dipoles that backscatter the incident wave in its orthogonal polarization. Each dipole is tuned at a predetermined frequency, and its presence or absence codes the corresponding bit state. The use of orthogonal polarization mitigates the clutter reflections and coupling between the reader's antennas, allowing a better detection of the tags.

Amplitude-phase backscatter modulation-based chipless RFID tags operate at narrower bandwidths compared with time- or frequency-coded tags. These types of tags encode the ID varying the amplitude or phase of the backscattered signal due to the load connected to an antenna. Left-hand (LH) delay line based tags consist of a narrowband antenna connected to a series of cascaded LH delay lines [SCH 09]. Each LH section produces a discontinuity in the phase of an incident wave. The reader interrogates the LH-based tag using a modulated signal, such as quadrature phase shift keying. Each tag produces a unique phase variation on the carrier signal. Remote complex impedance-based chipless tags [MUK 07] are formed by a printable scattering antenna (for instance a patch antenna) loaded with a lossless reactance. Each tag has a unique reactance that generates a unique inductive loading. The backscattered signal thus has a different phase for each tag. Stub-loaded patch antenna based tags [BAL 09a] are similar to remote complex impedance-based tags, with increased robustness. In this case, a stub loads a patch antenna. The ID is coded in the cross-polarized phase difference between electric (E) and magnetic (H) planes. Finally, carbon nanotube (CNT) loaded chipless tags consist of RFID antennas loaded by CNTs, which modify the scattering signature depending on their state. In [YAN 09], a conformal UHF RFID antenna is loaded with single-walled CNTs to realize a chipless RFID gas sensor.

In summary, chipless RFID is a field of interest in RFID. There is not a common standard as in passive UHF RFID. Therefore, chipless RFID is still the subject of active research. There are advantages and disadvantages between each approach, and the final application will decide which approach is chosen. Most of the published work on chipless RFID relies on using high-cost laboratory instruments as readers to demonstrate the feasibility of the proposed tags. However, there is an increasing interest in developing custom readers [TED 13], which would reduce costs and enable the adoption of chipless RFID for specific market niches.

1.2. Extending RFID capabilities: from ID to sensing

There have been many advances in the miniaturization and cost reduction of advanced sensors [YIC 08]. A large number of applications can benefit from the information about their environment these sensors can obtain. Smart homes [HEL 05] or smart cities [HEL 05] are concepts that have flourished recently. In both cases, one of the main ideas is to fill an area (either houses or cities) with small, self-autonomous and low-cost sensors. These sensors are connected creating so-called sensor networks [YIC 08]. For large-scale applications, wiring each sensor is not viable. Also, some sensors can be placed in areas difficult to access. Therefore, wireless radio technologies that enable the sensors to be read remotely, creating a sensor node, are desired. The association of these wirelessly readable sensor nodes is called a wireless sensor network (WSN). Apart from smart homes or smart cities, WSNs also have a great potential in a large number of applications such as [YIC 08] military target tracking and surveillance, natural disaster relief, biomedical health monitoring, hazardous environment exploration and seismic sensing. Low-power communication technologies are required to achieve years of lifetime for wireless sensors. Careful design for these technologies, based on small data rates (for small amounts of information, only sensor readings), has to be taken into account.

Sensor nodes and readers (or reading points (RP)) can be associated in several ways. A direct wireless link between each sensor and RP (see Figure 1.6(a), centralized "star" topology) is a solution that requires high RF power transmitters if distances are long, with a consequent impact on battery lifetime or the need for a power supply to each sensor. A second possibility is to link the sensors in what is called a wireless network. In this way, the sensors also act as a bridge for other sensors (see Figure 1.6(b), mesh topology). A third solution might be the use of a mobile link between sensors and RP (see Figure 1.6(c), mobile topology). This means taking advantage of the mobility of vehicles inside a city, by using them to also transport information. For instance (see Figure 1.6(d)), buses can be good candidates. A bus always performs the same route, and stops periodically and repeatedly in bus stops. If a wireless sensor is installed in a bus stop and the reader at the bus, the sensor can be read every time the bus stops there. The information recorded in each trajectory can be downloaded at a point (normally another bus stop), which is connected to the data management point. The main advantage of this solution is that low-power consumption wireless sensors can be used, since read ranges are in the order of few meters,

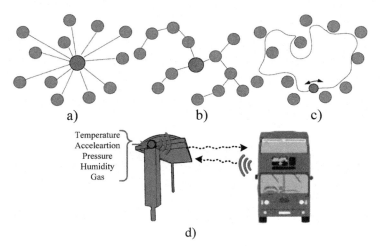

Figure 1.6. *Sensors: green points. Reader point (RP): red point.*
a) Centralized topology, b) mesh topology, c) mobile RP topology and
d) scheme of transmission between the bus and the bus station. For a
color version of the figure, see www.iste.co.uk/ramos/rfid.zip

1.2.1. *Existing technologies for WSNs*

There are several commercial low-power solutions on the market for WSNs [NAP 11, SMI 11, GUN 09, TEX 14, DEM 13]. Some of the most popular solutions are now compared in terms of cost, power consumption, speed and range. Table 1.1 summarizes typical values of these parameters for bluetooth low-energy (LE), ANT and ZigBee technologies. Other technologies such as Wi-Fi (IEEE 802.11), NFC or infrared (IrDA) [SMI 11] were discarded in this comparison because they are not intended for wireless sensor applications; either their power consumption is extremely high for miniaturized portable devices (Wi-Fi) or their read range is very short for a wireless sensor (NFC or IrDA).

– *Bluetooth LE*: although its original aim was for mobile devices and accessories, the latest specification (version 4.0 LE) in 2010 takes into account low-power devices. Wearable devices, such as cardiac sensors or pedometers, companions for smartphones, are enabled with bluetooth LE [WAN 13].

– *ANT*: a proprietary technology by Dynastream, working at 2.4 GHz as bluetooth. It is aimed towards wearable devices in combination with smartphones [DYN 15].

– *ZigBee*: a specification given by the IEEE 802.15.4 standard, which is specifically intended for home automation and larger areas than bluetooth LE or ANT [WHE 07].

Technology	Approximate cost per unit (US $)	Sensitivity (dBm)	Power consumption (μW/bit)	Peak current (mA)	Speed (kbps)	Approximate range (m)
Bluetooth LE	2.95	−87	0.153	12.5	305	100 @ + 0 dBm Tx
ANT	3.95	−85	0.71	17	20	30 @ + 0 dBm Tx
ZigBee	3.20	−100	185.9	40	205	100@ + 1 dBm Tx

Table 1.1. *Comparison between existing technologies for WSNs with typical values*

As can be observed, bluetooth LE has very low average power consumption; however, its peak current is very large for battery-less devices, which rely on external sources [PAR 05]. These external sources can be the reader's RF signal, solar energy or movement, for instance [KIM 13]. As introduced in section 1.1, passive RFID tags are powered from the reader's signal.

1.2.2. *RFID-enabled wireless sensors*

The RFID reader uses a wireless communication link when it retrieves the ID from one or several tags. This link can also be exploited to collect data from a sensor connected to or embedded into the tag [KIM 13, WAN 04]. Adding sensing capabilities to RFID permits possibilities beyond what barcode systems offer. In addition, RFID systems have less complicated protocols and data frames than ZigBee or Bluetooth, for instance. Also, even though ZigBee or bluetooth can offer faster absolute data rates, the communication with RFID is established faster because the tag does not need to associate and authenticate with the reader at the beginning. One typical application for RFID-enabled wireless sensors is monitoring the cold chain in perishable products. The customer, as well as the seller and logistics companies, can determine the temperature range of the item from its production to its final destination. Accelerometers can also be used in fragile products in order to detect hits or bad package handling. As an example, in [GON 14], an RFID-enabled sensor tag is embedded in cork wine bottle stoppers to monitor their temperature.

RFID systems also have a several advantages for WSNs in smart homes or smart cities applications. The cost of RFID tags can be very low when using low-cost substrates and inkjet printing technology [MOL 03]. The architecture of RFID systems is also simpler than other systems such as bluetooth LE or ZigBee (see section 1.2.1), because the sensor tags do not require dedicated transceivers. It is also possible to integrate RFID systems in conventional WSNs, as shown in [LIU 08]. RFID-enabled sensors are integrated with materials that are sensitive to physical parameters, for instance water-absorbing materials for humidity sensors and carbon nanostructures for gas sensors [VEN 13a]. The electrical parameters of the materials (such as permittivity and conductivity) are changed by the physical parameters. These electrical changes are translated in changes in the RFID signal. In the last few years, some platforms based on microcontrollers that emulate the behavior of passive UHF EPC Gen2 tags have been presented. The best known example is the wireless ID and sensing platform [SAM 08] from Intel Research Seattle. A photograph is shown in Figure 1.7. Other similar platforms based on inkjet printing on paper substrates have been presented [VYK 09].

Figure 1.7. *Photograph of the wireless identification and sensing platform (WISP) RFID-enabled sensing platform*

1.3. Ultra-wideband technology for RFID applications

1.3.1. *Introduction to ultra-wideband technology*

Ultra-wideband (UWB) radio technology uses very short (nanosecond order) time domain pulses [FCC 03, JOS 04]. Using these kinds of pulses widens the signal in the frequency domain to be much wider than traditional communications that use narrowband frequency-multiplexed signals. A UWB signal is defined as a signal with a bandwidth higher than 20% of its center frequency, or a signal with a bandwidth higher than 0.5 GHz.

The American Federal Communications Commission (FCC) specified a band of operation for UWB signals from 3.1 to 10.6 GHz in 2002 [FCC 03]. This band can be used freely, with the only limitation of radiated power. Therefore, UWB signals cannot affect traditional narrowband communications. In Europe, the European Telecommunications Standard Institute (ETSI) and the European Conference of Postal and Telecommunications Administrations specified a slightly different power mask for UWB communications [ETS 08, AST 09]. Figure 1.8 shows the maximum allowed indoor and outdoor levels by both ETSI and FCC. Evidently, European regulations are more restrictive than American regulations. The FCC also allowed operation of UWB in 2002 while Europe did in 2005. Since the ETSI does not regulate a single country as the FCC does, approving the standard is slower and the final mask is more restrictive, to comply with all countries' existing narrowband systems [MAZ 11]. From Figure 1.8, there is a band below 1 GHz that is intended for ground penetrating radar (GPR) systems. GPR systems point the radar antenna(s) toward the ground floor, and therefore are not likely to cause interference on other systems [DAN 05]. Figure 1.9 compares the transmitted power of a UWB signal and a narrowband signal as a function of frequency. It can be clearly seen that UWB signals require less power than narrowband signals, but their bandwidth is much higher than narrowband systems.

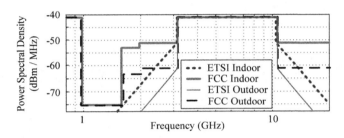

Figure 1.8. *Maximum indoor and outdoor levels of power spectral density (PSD) as a function of frequency for ETSI and FCC regulations. For a color version of the figure, see www.iste.co.uk/ramos/rfid.zip*

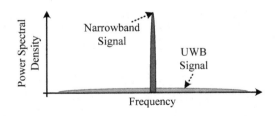

Figure 1.9. *Comparison between power levels and frequency bandwidth of a narrowband and a UWB signal*

Since UWB pulses have this large bandwidth, a UWB system permits better immunity to multipath propagation and narrowband interferences, because these kinds of interferences only affect a part of the complete spectrum. UWB also has good penetration in materials. Another advantage of UWB technology for RFID resides in the size of the antennas, which is usually smaller than traditional narrowband RFID due to the increase in the operating frequency. There exists rising demand for new antenna designs to have small dimensions. The higher frequencies of UWB enable compact hand readers and an increase in resolution position for localization systems. RF circuitry can be simpler with UWB, and data transfer rates can be higher. Therefore, an interest in UWB technology has arisen in industry and research fields [FON 04].

1.3.2. UWB-based RFID

UWB technology can be a promising solution for next-generation RFID systems due to the advantages inherent to its large bandwidth. Many frequency bands from 9 kHz to 24 GHz are theoretically capable of being used in RFID. Some of these allocated bands are denominated industrial, scientific and medical (ISM), and they are usually free to use without any license in many countries. In front of the UHF band for RFID, ISM bands, especially the most popular one at 2.4 GHz, is saturated because of wireless LAN and bluetooth applications, leading to a poor performance when using it for RFID. However, UWB presents a frequency band that is much higher than allocated narrowband frequencies.

Additionally, even though ISM bands do not need licenses, their emitting power is limited to avoid interferences, leading to a majority of active-only RFID tags in these bands, and the highest power consumption is precisely in the RF transmission part. UWB impulses need less power than narrowband signals, which means that UWB can be used to develop low-power active and semi-passive tags in the future. Moreover, UWB is able to resolve the growing demand for higher data transmission speeds. The bandwidth with UHF and ISM bands is usually not sufficient for the resolution required in indoor localization applications. One of the most important commercial applications of UWB (and GPR) is indoor localization due to their large bandwidth.

Despite all these potential advantages, it is still necessary to improve certain aspects such as cost reduction, tag tracking precision (regardless of its speed or read rate) and reading reliability. Recently, UWB-based chipless RFID systems have been proposed in the literature [BAL 09b, ZHA 06]. The number of IDs that chipless frequency-coded RFID can encode depends on the allowed bandwidth. Thus, the UWB is often used in these tags. In [BAL 09b], chipless printable RFID tags are proposed by using several resonators in frequency domain. In [SCH 09, DOW 09], sensors are integrated with chipless RFID tags to remotely read the sensor. In chipless passive RFID UWB tags, the read range is not limited by the power threshold to activate the chip, which is the main limitation for read distance of passive UHF tags [LAZ 09b].

Moreover, multipath interferences can positively or negatively affect the read when working with UHF RFID tags. The tag may not be readable even though it is inside the read range due to multipath [LAZ 09b]. This situation can be resolved using UWB technology, since different responses originated by the multipath interference can be minimized by using signal-windowing techniques in time domain.

An alternative method of using several resonators in the frequency domain consists of coding the information in the time delay [DOU 10, DAR 08]. Here, the simplest way to code information is by varying the physical length of an open-ended transmission line connected to a scattering UWB antenna. The length of the transmission line changes the time delay of the reflection due to the tag antenna, and therefore different states can be coded. Although this idea has been proposed by some authors [DOU 10, DAR 08], there are few experimental results, which have been obtained by means of high-cost instruments such as vector network analyzers. Future implementations of commercial readers should be based on low-cost equipment, such as impulse radio UWB radars.

1.4. Organization of this book

The book is organized as follows. Chapter 2 describes the chipless time-coded UWB RFID theory, signal processing techniques and reader alternatives. Several designs of chipless tags are shown and characterized. A study and discussion on read range, resolution, number of bits, influence of angle, polarization, materials and tag bending are presented. Chapter 3 uses foundations and tags shown in Chapter 2 to design chipless sensors. Amplitude-based (continuous and threshold temperature) and delay-based (permittivity for concrete composition detection) chipless sensors are presented as examples. Chapter 4 describes semi-passive sensing platforms based on time-coded UWB RFID. Two topologies based on analog and digital approaches are explained. Chapter 5 integrates sensors in the semi-passive sensing platforms from Chapter 4, and a temperature and a gas sensor (this later based on CNTs) are shown as examples. Chapter 6 shows a smart floor application where chipless (Chapter 2) and semi-passive (Chapter 4) tags are combined with GPR techniques. Finally, Chapter 7 presents active long-range platforms based on time-coded UWB RFID intended for localization applications.

Chipless Time-coded UWB RFID: Reader, Signal Processing and Tag Design

2.1. Introduction

Chipless time-coded UWB RFID could be an alternative for RFID systems. The tag's ID is coded in the physical length of an open-ended transmission line connected to a scattering UWB antenna. Although this idea has been proposed by some authors [HU 07, HU 08, HU 10, DAR 08], there are few experimental results, which have been obtained by means of expensive instruments such as vector network analyzers (VNAs). Future implementations of commercial readers should be based on low-cost equipment, such as impulse radio (IR) UWB radars [GEO 14, NOV 12, TIM 14]. Due to their large bandwidth, a small signal-to-noise ratio (SNR) is expected with these systems. Therefore, signal processing techniques should be considered in order to detect the tag in a real scenario with noise. Finally, the realization (integrating a UWB antenna with a long delay) and characterization of chipless time-coded UWB tags should be studied in detail. In this chapter, the following fields are addressed:

– section 2.2 presents the theoretical foundations;

– section 2.3 compares two reader approaches;

– section 2.4 presents the signal processing techniques used to improve the detection in real scenarios;

– section 2.5 studies the design of tags in terms of integration of UWB antennas and delay lines;

– section 2.6 characterizes the tags in terms of the materials attached, tag-reader angle, polarization and bending;

– finally, section 2.7 draws the conclusions of the chapter.

2.2. Theory

Passive RFID is based on modulating the radar cross-section (RCS) [RAO 05] of the tag. Following various studies [GRE 66, COL 69, HAN 89], there are different formulations for deriving the scattered field at an antenna connected to an arbitrary load, when the antenna is illuminated by a plane wave. However, these studies have shown that this field can be expressed as the sum of two terms (or modes):

– a structural mode E^{sm}, which is mainly due to the wave diffraction at the antenna structure (patches, ground plane, edge effects, etc.);

– an antenna mode (or tag mode) E^{am}, which is mainly due to the radiation properties of the antenna. This term depends on the load Z_L connected to the antenna.

As a consequence, the scattered field $E^S(Z_L)$ at an antenna connected to an arbitrary load Z_L can be obtained from [LIU 03, JOH 82] and expressed as:

$$\overline{E^S}(Z_L) = \overline{E^{sm}}(Z_c) + \overline{E^{am}}(Z_L) = \overline{E^{sm}}(Z_c) + \frac{\Gamma_L}{1-\Gamma_L\Gamma_a}\overline{E_0}, \qquad [2.1]$$

where $E^{sm}(Z_c)$ is the structural mode scattering field and $E^{am}(Z_L) = E_0\Gamma_L/(1-\Gamma_L\Gamma_a)$ is the antenna mode (or tag mode) scattering field, Z_c is the normalization impedance, E_0 is the scattering field under a unit incident wave, and Γ_a and Γ_L are the reflection coefficients of the antenna and the load, respectively. The reflection coefficient Γ_L, which multiplies the unit incident wave scattering field E_0, depends on the circuit connected to the antenna. This circuit not only accounts for the load itself (Z_{LOAD}), but also for the transmission line that connects the antenna and the load. Therefore, the antenna mode scattering field depends on the load and length L of this transmission line. When the circuit connected to the antenna is matched ($\Gamma_L = 0$), only structural scattering exists. If not, part of the received energy is reradiated and structural and antenna modes coexist.

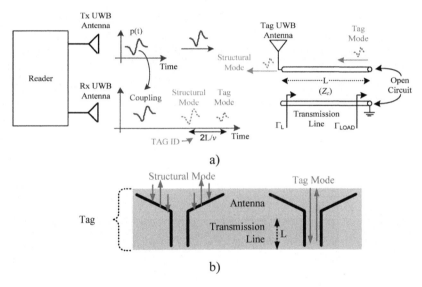

Figure 2.1. *a) Time-coded UWB RFID system scheme; b) scheme of the structural and tag modes. For a color version of the figure, see www.iste.co.uk/ramos/rfid.zip*

The full RFID scheme is shown in Figure 2.1(a), and a scheme of the structural and tag modes is shown in Figure 2.1(b). The transmitter (Tx) illuminates the tag. When the transmitted pulse hits the tag antenna, a portion is backscattered toward the receiver (Rx) and a portion propagates inside the tag. Then, time-coded chipless tags can be considered scattering antennas (antennas terminated with load impedance) with two scattering modes: the structural mode (first or early-time reflection) and the tag (or antenna) mode (second reflection). Also, coupling from Tx antenna to Rx antenna has to be considered, since both antennas cannot be perfectly isolated from each other in a real case scenario.

The tag is modeled as an equivalent two-port network (antenna) terminated with a transmission line of length L and characteristic impedance Z_c. The line is in turn loaded with impedance Z_{LOAD} [COL 69], as presented in Figure 2.2. The wave a_{in} represents the incoming wave from the reader. The out coming wave b_{out} is generated due to reflection and is scattered in direction to the reader. Wave c_{pl} represents the coupling from the reader transmitting to receiving antennas. The waves a_{in} and b_{out} are normalized to the free-space impedance ($120\pi\ \Omega$). The output of the antenna is normalized to Z_c. Thus, S_{22a} in Figure 2.2(b) represents the reflection coefficient of the antenna, $S_{22a} = \Gamma_a$. The reflection coefficients Γ_a and Γ_L are defined as [2.2] and [2.3], respectively:

$$\Gamma_a = \frac{Z_a - Z_c}{Z_a - Z_c} \qquad\qquad [2.2]$$

$$\Gamma_L = \frac{Z_L - Z_c}{Z_L - Z_c}, \qquad\qquad [2.3]$$

where Z_a is the antenna impedance and Z_L the load connected to the antenna. The reflection coefficient at the input of the tag Γ_{in} can be obtained from the analysis of Figure 2.2(b):

$$\Gamma_{in} = \frac{b}{a} = S_{11a} + \frac{\Gamma_L}{1 - \Gamma_a \Gamma_L} S_{21a} S_{12a}, \qquad\qquad [2.4]$$

which can be expanded in series:

$$\Gamma_{in} = S_{11a} + S_{21a} S_{12a} \Gamma_L \left[1 + \sum_{n=1}^{\infty} (\Gamma_a \Gamma_L)^n \right] \approx S_{11a} + S_{21a} S_{12a} \Gamma_L. \quad [2.5]$$

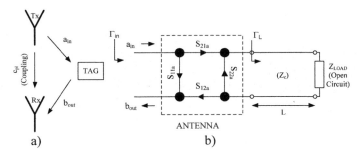

Figure 2.2. *Model for the UWB RFID tag*

As observed in [2.5], an approximation has been considered. S_{11a} represents the reflection in the antenna (structural mode), whereas S_{21a} $S_{12a} \Gamma_L$ represents the reflection in the load. Assuming that the RFID reader transmits a pulse $p(t)$ and defining $\tau_L = 2L/v$ as the round-trip propagation delay along the transmission line (v is the propagation velocity in the transmission line), a physical interpretation of [2.5] can be obtained from the bounce diagram shown in Figure 2.3. It is a two-dimensional representation of the transient waves bouncing back and forth on the tag. Zigzagging lines indicate the progress of the wave as a function of position and time. The direction of travel is from bottom to top. The terms within the series in [2.5] represent the multiple reflections of the waves between the load Z_{LOAD} and the antenna, which appear every time delay $n\tau_L$ (for $n = 1, 2, 3$, etc.). τ_p is the round-trip time delay between the tag and the reader, and τ_A is the delay introduced by the antenna itself. For a well-matched antenna, only the first term is considered because the others vanish rapidly. Since delay information in this term is the key parameter, the best method to obtain the maximum amplitude is to make $Z_{LOAD} = \infty$ or $Z_{LOAD} = 0$ (open-circuit or short-circuit load, respectively), and then design a line with Z_c matched to the antenna input impedance Z_a. Assuming a low-loss line, the reflection coefficient Γ_L is:

$$\Gamma_L = \Gamma_{LOAD} e^{-j2\pi f 2L/v} = \Gamma_{LOAD} e^{-j2\pi f \tau_L} , \qquad [2.6]$$

where f is the operating frequency and Γ_{LOAD} is the reflection coefficient of the load connected at the end of the transmission line (e.g. $\Gamma_{\text{LOAD}} = 1$ when $Z_{\text{LOAD}} = \infty$). It can also be seen that the phase of Γ_L, $e^{-j2\pi f 2L/v}$, directly depends on the frequency and increases with the length L. This shows that the load reflection coefficient phase, and therefore the scattered tag (antenna) mode field $E^{\text{am}}(Z_L)$, depends on the length of the transmission line.

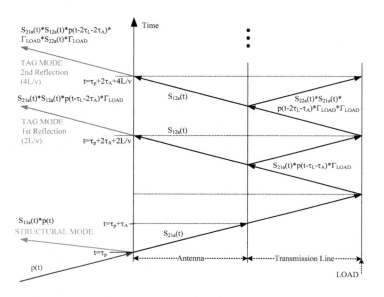

Figure 2.3. *Bounce diagram for transient waves scattered at the tag. For a color version of the figure, see www.iste.co.uk/ramos/rfid.zip*

By applying the inverse Fourier transform to [2.5], we can obtain the time-domain backscattered field or, equivalently, the time-domain reflection coefficient between the incoming and outcoming waves:

$$\Gamma_{in}(t) \approx S_{11a}(t) + S_{12a}(t) * S_{21a}(t) * \delta\left(t - \tau_L - 2\tau_A\right) * \mathfrak{I}^{-1}\left(\Gamma_{\text{LOAD}}\right), \quad [2.7]$$

where $*$ denotes the convolution operator. When Γ_{LOAD} is real (i.e. resistive or open-/short-circuited loads), [2.7] can be expressed as:

$$\Gamma_{in}(t) \approx S_{11a}(t) + \Gamma_{\text{LOAD}} g\left(t - \tau_L - 2\tau_A\right), \quad [2.8]$$

where $g(t)$ is defined as the inverse Fourier transform of $S_{12a}S_{21a}$:

$$g(t) = \Im^{-1}\left(S_{12a}(f)S_{21a}(f)\right) = S_{12a}(t) * S_{21a}(t).$$ [2.9]

Since the structural mode $S_{11a}(t)$ and $g(t)$ have a finite time duration, the time responses associated with the structural mode and the tag mode $g(t-\tau_L)$ can be separated if the line length L is conveniently designed. The received signal at the reader in the frequency domain is then given by:

$$S(f) = H_{\text{free}}(f,r_1)\Gamma_{\text{in}}(f)H_{\text{free}}(f,r_2)P(f).$$ [2.10]

$P(f)$ is the Fourier transform of the transmitted pulse $p(t)$ (which includes the response of the transmitting antenna), r_1 is the distance from the reader's transmitting antenna to the tag and r_2 is the distance from the tag-to-reader's receiving antenna. H_{free} is the transfer function due to free space propagation:

$$H_{\text{free}}(f,r) = \frac{1}{\sqrt{4\pi \cdot r}}e^{-j2\pi f \cdot r/c} \xrightarrow{\Im^{-1}} h_{\text{free}}(t,r) = \frac{1}{\sqrt{4\pi \cdot r}}\delta(t - r/c),$$ [2.11]

where c is propagation velocity in free space ($c = 3 \times 10^8$ m/s), $r = r_1 + r_2$ and $\delta(t)$ is the Dirac delta function. The term r/c represents the delay from the antenna, and the term $1/r$ represents the attenuation of a spherical wave. By applying the inverse Fourier transform to [2.10], the signal received at the reader in the time domain is given by:

$$s(t) = \Gamma_{\text{in}}(t) * h_{\text{free}}(t,r_1) * h_{\text{free}}(t,r_2) * p(t),$$ [2.12]

which can be expressed as:

$$s(t) = \alpha p(t - \tau_p) * S_{11a}(t) + \alpha\Gamma_{\text{LOAD}}p(t - \tau_p) * g(t - \tau_L - 2\tau_A)$$
$$= \alpha \cdot S_{11a}(t) * p(t) * \delta(t - \tau_L) + \alpha\Gamma_{\text{LOAD}}g(t) * p(t) * \delta(t - \tau_p - \tau_L - 2\tau_A),$$ [2.13]

where α is the round-trip attenuation factor due to propagation in free space. Both parameters are functions of the tag-reader distance r. In [2.13], $S_{11a}(t)*p(t)$ is the response associated with the structural mode and $g(t)*p(t)$ is the response associated with the tag mode.

In the ideal case of an antenna with infinite bandwidth, $S_{11a}(t)$ and $g(t)$ can be approximated by the Dirac delta function, $\delta(t)$. Then, time resolution only depends on the type of pulse. However, the finite time duration of structural and tag modes produce an increase in the received pulse duration and some shape distortion, reducing the time resolution. This resolution determines the minimum delay that can be coded and, at the end, the number of data bits available. It will be experimentally studied in section 2.6.1. For each tag, the ID is coded in the time difference between the structural and tag modes:

$$\text{ID} = \left[\left(\tau_p \right)_{\text{Str.}} - \left(\tau_p - \tau_L - 2\tau_A \right)_{\text{Tag}} \right] = \tau_L + 2\tau_A,$$ [2.14]

where Str. accounts for the structural mode and Tag accounts for the tag mode. Between two tags with different ID (different lengths L), the delay introduced by the antenna $2\tau_A$ is the same. Therefore, $2\tau_A$ can be considered as an offset term applicable to all tags.

In real measurements, the tag response is distorted by clutter and cross-coupling between the reader's antennas. Clutter is defined as those scattering contributions not originated at the object under test (for instance reflections at the walls or other objects). Then, the signal received at the reader can be expressed as:

$$s'(t) = s(t) + s_c(t) + s_m(t),$$ [2.15]

where $s_c(t)$ is coupling contribution and $s_m(t)$ is the clutter due to multipath reflections.

As seen in Figure 2.3, the waves that finally return to the reader are those due to the structural mode and the tag (antenna) modes. The tag modes repeat themselves every time instant nL/v (where $n = 2, 4, 6$, etc.). In practice, only the first tag mode, which corresponds to the delay of $2L/v$, has enough amplitude to be detected. This means that the multiple tag modes, even though they exist theoretically, do not appear, and only the structural mode and the first tag mode are visible. Nevertheless, when the tag mode is heavily amplified by an active non-chipless tag, these multiple tag modes can be detected, as will be shown in Chapter 7.

2.3. Reader design

Two techniques are used to measure the tags. These techniques differ in the stimulus signal applied to the transmitting antenna of the reader. Next, both techniques are explained in detail, comparing each of them in terms of cost, accuracy and speed.

2.3.1. *Frequency-step approach*

This approach is based on using a VNA. In this particular case, it corresponds to an Agilent PNA E8364C [KEY 14]. The VNA is connected to a control PC with a GPIB bus, and calibrated at the antennas' reference plane, with 0 dBm of output power and with 1601 points, between 1 and 10 GHz. The reader's transmitting antenna (Tx) is connected to the VNA port 1. Similarly, the receiving antenna (Rx) is connected to the VNA port 2. The reader's antennas consist of a Vivaldi antenna (a scaled version of the Vivaldi antenna in [LAZ 11a]), with a good matching over the UWB band. Figure 2.4(a) shows the measured $|S_{11}|$ parameter.

When using a VNA, a sine wave is frequency stepped or continuously swept over the band of interest. The time-domain response is obtained from the inverse Fourier transform of the scattering parameter S_{21}. The step width determines the unambiguous range. Attention should also be paid to the inverse Fourier transform if the stimulus band is smaller than the antenna bandwidth. In this case, the side lobes in the impulse response are no longer determined by the antenna response but rather by the measurement bandwidth. These side lobes can be suppressed by windowing the data before the transformation, but doing this results in slightly reducing the range resolution (compared with using a rectangular window) [AGI 12]. In this case, a Hamming window has been chosen. Finally, the inverse Chirp-Z transform has been used as an alternative to the inverse Fourier transform to reduce computational time [MEY 06]. Figure 2.4(b) shows a scheme of the experimental setup using a VNA, and Figure 2.4(c) shows a photograph of the VNA, indicating the ports where the transmitting and receiving antennas are connected.

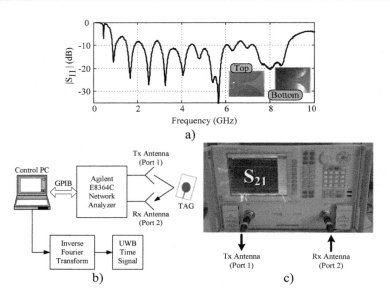

Figure 2.4. a) Simulated $|S_{11}|$ of the antennas used in
the reader. In the inset, photographs of the antennas;
b) scheme and c) photograph of the frequency-step approach

2.3.2. *Impulse-based approach*

In this approach, a UWB pulse is transmitted by the reader's transmitting antenna. In order to concentrate the energy of the stimulus in the pass band of the antenna, and therefore the UWB band, monocycles are used rather than pulses. The reflected signal at the tag $s(t)$ can be measured by means of an oscilloscope or a fast sampler. A scheme of the impulse-based approach is shown in Figure 2.5. Three commercial IR-UWB radars from different manufacturers (Geozondas, Novelda and Time Domain) are used as readers, and their specifications are depicted in Table 2.1.

The radar transmits the pulse using a pulse generator. A fast sampler, which triggers the pulse generator, receives the signal backscattered at the tag. Photographs of each radar are shown in Figure 2.6. In the Geozondas radar, the sampler and the pulse generator are two separated blocks and are externally connected by a trigger cable. On the contrary, the Novelda and Time Domain radars have them integrated in the same block. In the case of the Novelda radar, a monolithic integrated circuit includes the pulse generator and the sampler. In the case of the Time Domain radar, the transmitter and receiver are integrated, but the sampler is made by an external field programmable gate array. In order to sample the received signal, the Geozondas radar uses a sampling oscilloscope, the Novelda radar uses a method based on a continuous time binary valued (CTBV) design paradigm [TAY 12] and the Time Domain radar uses a rake receiver [PET 12]. The power consumption of the integrated radars is lower than the discrete one.

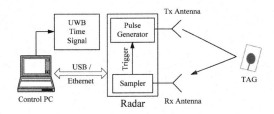

Figure 2.5. *Scheme of the impulse-based approach*

Radar	Resolution (ps/point)	Maximum required power	Number of points	Time window (ns)	Central frequency (GHz)	Measured band width (GHz)	Sampling method
Novelda NVA6100 (R640 Dev. Kit) [NOV 12]	30	120 mW	512	15.36	4.35	1.5	CTBV
Geozondas GZ6E (sampler) + GZ1120ME-35EV (pulse generator) [GEO 14]	25	15.6 W (1.3 A @ 12 V)	4,096	20	3.5	1.9	Sampling oscilloscope
Time Domain PulsON P400 MRM [TIM 14]	61	6.90 W	480	29.28	4.3	1.1	Rake

*Approximated at −3 dB from maximum amplitude (see Figure 2.7(b)).

Table 2.1. *Specifications of the commercial radars used for the impulse-based approach*

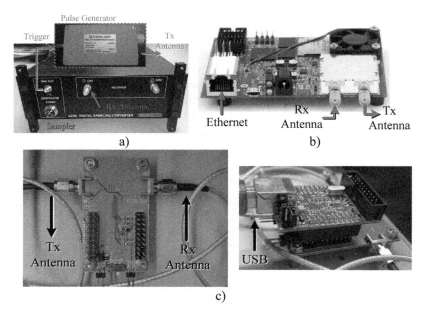

Figure 2.6. *Photograph of the a) Geozondas, b) Time Domain and c) Novelda radars used as time-coded UWB RFID readers. For a color version of the figure, see www.iste.co.uk/ramos/rfid.zip*

Figure 2.7 shows a comparison of the pulses reflected at a large metal plate using the Vivaldi antennas as shown in Figure 2.4(a). Depending on the technique used to generate the pulse [FON 04], its shape is different in time domain. Moreover, the antennas introduce a derivative effect that distorts its shape. However, as it can be seen in frequency domain, the NVA6100 and P400 pulses mostly comply with FCC regulations, since they are centered at 4.3 GHz. The pulse from the GZ1120ME-35EV generator is centered at a lower frequency and therefore a part of it falls outside the FCC mask. However, as will be seen in section 3.2.2, it can be used for certain sensing applications where the sensor performance is limited in frequency.

2.3.3. *Comparison and conclusions*

The advantages of the step-frequency approach are its excellent drift stability and random noise suppression because of the narrowband receivers, as well as its flexibility in the choice of the stimulus band. It is, however, the most expensive and slowest method.

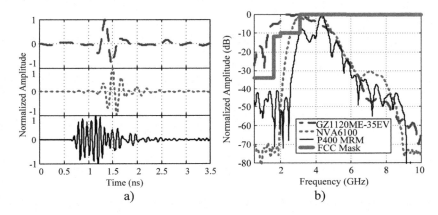

Figure 2.7. *a) Time and b) frequency domain response of the radar pulses reflected on a metal plate with Vivaldi antennas. For a color version of the figure, see www.iste.co.uk/ramos/rfid.zip*

The step-frequency approach also requires more computational power since it needs to apply the inverse Fourier transform each time a measurement is made. Thus, the measurement time of a VNA is several times larger than a IR-UWB radar and its power consumption is also greater than that of a radar. The step-frequency approach is a good method to take as reference.

The impulse-based approach, especially with integrated radars, provides a potentially low-cost, fast acquisition and portable solution. There are also commercial radar ICs that can be used to build fully custom readers [XET 15]. However, the resolution is limited as is the stability both in amplitude (offset) and time (jitter). In addition, the large bandwidth of the impulse-based approach reduces the SNR of the system. Table 2.2 summarizes the main differences between both approaches.

Method	Resolution	Power consumption	Price	Portable	Measurement speed	Computational requirements
Step-frequency	Large	Large	Very high	Maybe	Very slow	Large
Impulse-based (discrete)	Small–medium	Medium	Medium	Maybe	Fast	Small
Impulse-based (integrated)	Small	Small	Low	Yes	Fast	Small

Table 2.2. *Comparison between the step-frequency and impulse-based approaches*

2.4. Signal processing techniques

In a real-case scenario, RFID tags are read surrounded by other objects. These objects create reflections in the time-domain response. Moreover, walls and floors in the scene also generate large reflections, since their size is very large (meters) in comparison with a small tag (several centimeters). This is shown in the scheme of Figure 2.8. These reflections translate into undesired contributions, addressed as clutter, in the time-domain signal, which can distort or mask the actual tag response.

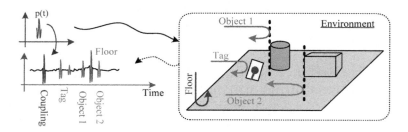

Figure 2.8. *Scheme of the time-domain signal with coupling and clutter reflections. For a color version of the figure, see www.iste.co.uk/ramos/rfid.zip*

The mean energy of a UWB pulse is very low, even for relative high amplitudes, and the noise bandwidth of the sampling converters is very large. As a consequence, the systems are sensitive to random noise. Noise influence may be reduced either by averaging (i.e. further reducing the measurement rate) or by generating extremely high voltage impulses. However, these possible solutions are not practical because of frequency regulations. Another drawback arises in the inadequateness of the sampling gate control due to nonlinearity in the ramp control, temperature drift or jitter. Possibly, UWB radars in the near future will overcome these drawbacks.

In order to overcome these limitations, the following section discusses several signal processing techniques. These techniques are intended to improve the quality of the received signal (SNR) after it has been received at the reader.

2.4.1. *Time-windowing and background subtraction*

Coupling between the readers' transmitting and receiving antennas can easily be neglected because it appears as a peak always at the same instant, regardless of the tag distance. It depends on the separation between the reader's antennas. It is also usually placed before the structural and tag modes, so by adding a delay before measuring, that corresponds to the time instant just after the coupling peak, coupling can be removed. This technique is known as time-domain windowing or gating [AGI 12]. The coupling and time response of the antennas imposes a minimum read distance or blind distance. This blind distance is lower than 10 cm for the setups used in section 2.3. In practice, it is not a problem because the systems are working on far-field.

Assuming that clutter and reader coupling are stationary, their effect is reduced by subtracting the empty room response (background) from the response in presence of the object. This process is called background subtraction. All contributions that do not overlap with the object response are ignored.

The procedure consists of first applying time-domain windowing to remove coupling and far echoes that are not in the time window of interest (where the tag is expected). Then, the background signal is measured and subtracted from the tag measurements.

An example of background subtraction is shown in the following. A raw food (unprocessed) measured signal for a time-coded UWB tag, using the setup from section 2.3.2 and the Geozondas radar, is plotted in Figure 2.9(a). Then, after applying background subtraction, the resulting signal is shown in Figure 2.9(b). It can be clearly seen that both structural and tag modes, which were not visible, can be distinguished.

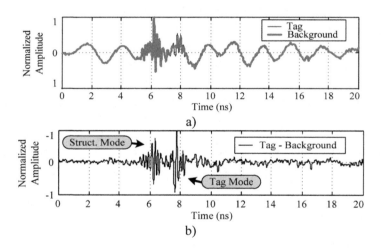

a)

b)

Figure 2.9. *a) Unprocessed RAW signal from a time-coded chipless tag and background scene; b) resulting signal from subtracting the background from the tag response. For a color version of the figure, see www.iste.co.uk/ramos/rfid.zip*

2.4.2. Continuous wavelet transform

This section explains in detail the continuous wavelet transform (CWT) [POU 03] as a detection technique for time-coded UWB RFID.

The "analytic signal" $s^+(t)$ is obtained from applying the Hilbert transform to the received signal $s(t)$ at the reader [POU 03]:

$$s^+(t) = s(t) + j\left(H\left[s(t)\right]\right), \qquad [2.16]$$

where $H[s(t)]$ is the Hilbert transform of the received signal $s(t)$, and j is the complex imaginary number (i.e. $j = \sqrt{-1}$).

The Hilbert transform is useful for obtaining the instantaneous envelope and frequency of a time series. The instantaneous envelope is the amplitude of the complex Hilbert transform (the complex Hilbert transform is the analytic signal) and the instantaneous frequency is the time rate of change in the instantaneous phase angle.

Figure 2.10 shows the previous background-subtracted signal from Figure 2.9(b) after applying the Hilbert transform. Since time-domain UWB RFID relies on the time difference between the structural and tag modes, there is no need to keep the shape of the modes, only the time position of the peaks is needed. Also, there is no need to know whether the signal is positive or negative. Therefore, the Hilbert transform obtains the envelope and removes both the sign and the shape information. By comparing both figures, it can be seen that the structural and tag mode peaks can be better localized when the Hilbert transform is applied. Although the Hilbert transform has noticeably improved the signal, it still has some noise that should be corrected in order to be able to detect in scenarios with more noise. This is done by means of the CWT.

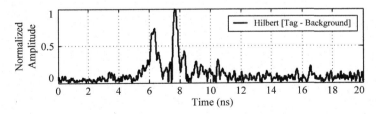

Figure 2.10. *Hilbert transform of the background-subtracted time-domain signal from Figure 2.9*

The CWT of a given function $s(t)$ (assuming it has zero mean and finite energy) is defined as a convolution [KAI 94]:

$$s(a,\tau) = \int_{-\infty}^{+\infty} s(t) \frac{1}{\sqrt{a}} \psi^* \left(\frac{\tau - t}{a} \right) dt , \qquad [2.17]$$

where * denotes complex conjugation. $s(a,\tau)$ is the wavelet coefficient at delay τ and scale a. The term $\psi^* \left(\tau - \frac{t}{a} \right)$ is the conjugated complex Gaussian wavelet, which is detailed in the following. This integral measures the comparison of the local shape of the signal and the shape of the wavelet. The dilation factor a is a measure of the duration of the event being examined, and by changing its value the signal can be zoomed in and out. Localization in time is achieved by selecting τ. Thus, time and frequency localization is achieved for each pair (a,τ) in the wavelet half-plane. The wavelets can be normalized to have constant energy at all scales. The convolution operation can be done in the Fourier domain as a product of the Fourier transforms, and then more efficiently implement the CWT by using the inverse fast Fourier transform algorithm [KAI 94].

The CWT can be interpreted as a matched filter to the received signal. From Schwarz's inequality, it is found that:

$$|s(a,\tau)|^2 = \left| \int_{-\infty}^{+\infty} s(t) \frac{1}{\sqrt{a}} \psi^* \left(\frac{\tau - t}{a} \right) dt \right|^2 \le \int_{-\infty}^{+\infty} |s(t)|^2 dt \int_{-\infty}^{+\infty} \frac{1}{a} \left| \psi^* \left(\frac{\tau - t}{a} \right) \right|^2 dt . \qquad [2.18]$$

The identity is held when $s(t)=k\psi((\tau-t)/a)$, where k is an arbitrary constant. It can be deduced from [2.18] that the amplitude of the CWT is maximized when the received signal shape is equal to one of the mother functions for a given optimum scale a_m and delay τ_m. In consequence, the arguments of the CWT a_m and τ_m specify the dilation and translation or delay, which characterize the received pulse. The mathematical expression for the maximum CWT amplitude arguments a_m and τ_m is written as:

$$(a_m, \tau_m) = \arg \left[\max (a, \tau) \right] \{ |s(a, \tau)| \} . \qquad [2.19]$$

It is known that a matched filter optimizes the SNR if the input noise is Gaussian. In UWB RFID, the transmitted waveform is, approximately, a Gaussian monocycle. Due to the derivative effect of the transmitting and receiving antennas, the received waveform is close to the second or third derivative of the Gaussian pulse (depending on the antennas used). If a mother family close to the expected waveform pulse is chosen, the CWT coefficients are the output of a matched filter or a correlator. Depending on the application, one wavelet family or another can be used [KAI 94]. Here, complex Gaussian wavelets have been selected, since Gaussian-like shaped pulses or their derivatives are used. The nth order complex Gaussian wavelet is obtained from the nth derivative of the complex Gaussian function $\psi_n(t)$:

$$\psi_n(t) = c_n \frac{d^n}{dt^n}\left(e^{-jt}e^{-t^2}\right),$$
[2.20]

where c_n is a normalization constant such that the two-norm of $\psi_n(t)$ is also. If n is even, the real part of ψ_n is an even function and the imaginary part is odd, and vice versa for an odd n. Therefore, the real and imaginary parts of the wavelet are orthogonal.

Instead of using the received signal $s(t)$, the analytical signal $s^+(t)$ obtained in [2.16] can be used [LAZ 09]. For instance, Figure 2.11(a) shows a third-order Gaussian pulse with central frequency 5 GHz, its Hilbert transform $H[s(t)]$ and its envelope. Figure 2.11(b) shows all the calculated CWT coefficients in a scaled picture (redder color means higher amplitude) and Figure 2.11(c) shows a cut for the maximum scale.

Figure 2.12 shows the estimated error in the time-of-arrival (ToA) of the pulse in presence of noise. The theoretical value is 0 ns, as shown in Figure 2.11. In this example, Gaussian noise is added to the received pulse to emulate a noisy measurement. The error is estimated from the peak of the envelope of the signal. To calculate the envelope, the Hilbert transform is applied to the signal and then the magnitude is obtained. After, the maximum of the CWT of the magnitude of the Hilbert transform of the signal is calculated, as shown in

Figure 2.11(c). This result shows the robustness of the CWT when there is presence of noise, since the error when using the CWT, marked with blue color, is much lower than when using only the Hilbert transform, which is marked with green-dashed color. The envelope-only technique also shows a higher error even considering the worst case for CWT (with a SNR of −10 dB) and the best case for Hilbert only (with a SNR of +30 dB).

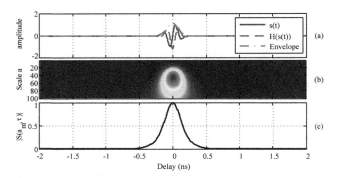

Figure 2.11. *a) Simulated time signal s(t), its Hilbert transform H(s(t)) and the envelope of the analytical signal s⁺(t), b) magnitude of the CWT of s⁺(t) using the third-order complex Gaussian wavelet, and c) cut of CWT for the scale of the peak magnitude. For a color version of the figure, see www.iste.co.uk/ ramos/rfid.zip*

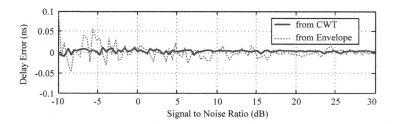

Figure 2.12. *Error in the ToA detected from the peak of the magnitude of the CWT (blue) and from the peak of the magnitude of the Hilbert transform (green dashed) as a function of the signal-to-noise ratio (SNR) in dB. For a color version of the figure, see www.iste.co.uk/ramos/rfid.zip*

Finally, Figure 2.13 shows a more realistic case. By applying the CWT to the signal from Figure 2.10, we can remove noise and improve the quality of the signal. The scale range used is between 1 and 64 with a third-order complex Gaussian wavelet. These parameters are kept for all the measurements shown in the following chapters.

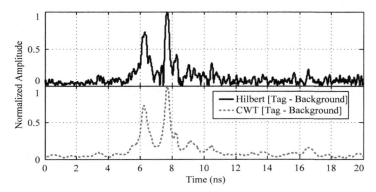

Figure 2.13. *Continuous wavelet transform of the background-subtracted time-domain signal from Figure 2.9. For a color version of the figure, see www.iste.co.uk/ramos/rfid.zip*

2.5. Design of chipless time-coded UWB RFID tags

As explained in section 2.2, chipless time-coded UWB RFID tags are composed of a UWB antenna connected to an open-ended delay line. It is desired that small, planar UWB antennas are integrated with long, compact delay lines in the same substrate. In this manner, small tags with a large number of possible IDs can be obtained. Moreover, a large-scale fabrication of tags can be of very low cost. No parts have to be soldered and they can be fabricated with additive technologies such as inkjet printing [SHA 11]. This section studies in detail the possibilities and limitations of integrating UWB antennas with delay lines on the same substrate.

2.5.1. *Design of UWB antennas*

Several UWB antennas are studied in the following for chipless time-coded RFID tags. They differ in the radiation pattern shape, size and gain. All of the antennas are fabricated on Rogers RO4003C substrate. The main properties, according to the manufacturer, are depicted in Table 2.3. Design and simulations are done with a Keysight Momentum Electromagnetic Simulator.

Property	Typical value
Dielectric constant, ε_r	3.55
Dissipation factor, tanδ	0.0027 @ 10 GHz, 23 °C
Substrate thickness	0.813 mm
Copper thickness	35 μm
Copper conductivity	4.1×10^7 S/m

Table 2.3. *Substrate properties for Rogers RO4003C substrate*

The first antenna is a broadband eccentric annular monopole UWB antenna [ANG 06]. Dimensions and photographs of the antenna are shown in Figure 2.14. The radiation element consists of a circular slot of radius 35 mm that is fed by a circular open-ended microstrip line of radius 8.97 mm. This line is connected to a 50 Ω access line. Figure 2.15 shows the simulated and measured $|S_{11}|$, the simulated maximum gain and the simulated gain at 4.3 GHz in H-plane. As explained in section 2.3.2, the center frequency of the Novelda and Time Domain radars is around 4.3 GHz. It can be observed that the antenna works correctly over the entire UWB band (3.1–10.6 GHz), with a $|S_{11}|$ under −10 dB. The maximum simulated gain at 4.3 GHz is 5.15 dB at 0° and 180° (symmetric) in H-plane.

The second antenna is a bow-tie antenna. Figure 2.16 shows the layout and photographs of the antenna. It consists of a tapered slot transition fed from a 50 Ω access line to a bow-tie shaped radiator in the bottom face. Figure 2.17 shows the simulated and measured $|S_{11}|$, and the simulated gain at 4.3 GHz in the H-plane. Again, a good agreement is observed, with a maximum gain of 2.97 dB at 0° and 180° (symmetric).

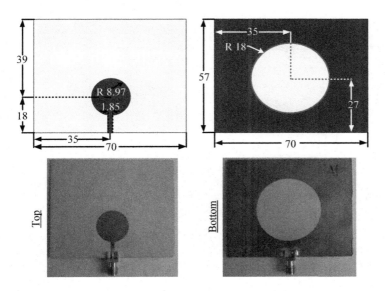

Figure 2.14. *Layout and photographs of the broadband eccentric annular monopole antenna. Dimensions are in millimeters. For a color version of the figure, see www.iste.co.uk/ramos/rfid.zip*

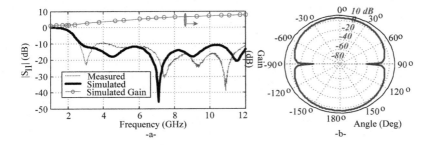

Figure 2.15. *a) Simulated and measured $|S_{11}|$, and simulated maximum gain of the broadband eccentric annular monopole antenna; b) simulated gain at 4.3 GHz in H-plane. For a color version of the figure, see www.iste.co.uk/ramos/rfid.zip*

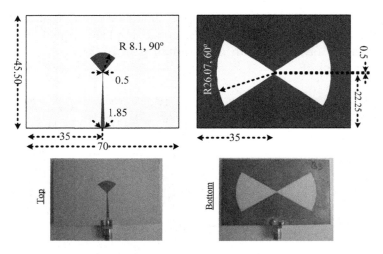

Figure 2.16. *Layout and photographs of the bow-tie antenna. Dimensions are in millimeters. For a color version of the figure, see www.iste.co.uk/ramos/rfid.zip*

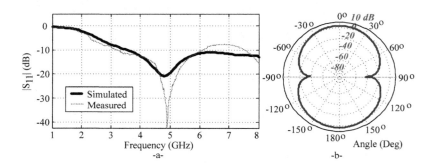

Figure 2.17. *a) Simulated and measured $|S_{11}|$ of the bow-tie antenna; b) simulated gain at 4.3 GHz in H-plane. For a color version of the figure, see www.iste.co.uk/ramos/rfid.zip*

The third antenna is a dipole antenna. Its layout is shown in the inset of Figure 2.18. It is composed of two ellipses and an access line. As shown in the simulations from Figure 2.18, the antenna is well matched for frequencies above 1.2 GHz. The maximum gain is achieved at −111° with 3.12 dB.

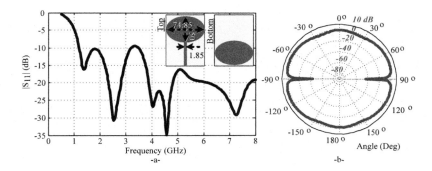

Figure 2.18. *a) Simulated |S₁₁| of the dipole antenna; b) simulated gain at 4.3 GHz in H-plane. In the inset, layout and dimensions of the antenna. For a color version of the figure, see www.iste.co.uk/ramos/rfid.zip*

Finally, the last antenna studied is a Vivaldi antenna. Figure 2.19 shows the layout of the antenna with its dimensions. It consists of a tapered slot transition with an exponential taper radiator on the bottom face. Figure 2.20 shows the simulated $|S_{11}|$ and the simulated gain in H-plane at 4.3 GHz. A good matching for frequencies above 1.5 GHz is observed. In this case, the shape of the radiation pattern is noticeably different from the other antennas, with a maximum radiation lobe at 68°. The maximum gain at this angle is 6.52 dB, which is larger than in the other designs.

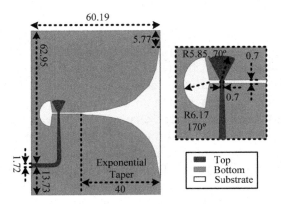

Figure 2.19. *Layout of the tapered slot Vivaldi antenna. Dimensions are in millimeters. For a color version of the figure, see www.iste.co.uk/ramos/rfid.zip*

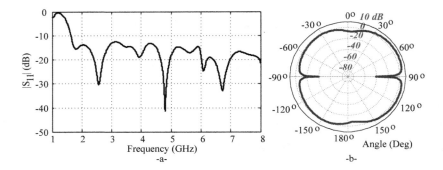

Figure 2.20. *a) Simulated $|S_{11}|$ of the Vivaldi antenna; b) simulated gain at 4.3 GHz in H-plane. For a color version of the figure, see www.iste.co.uk/ramos/rfid.zip*

2.5.2. *Integrating delay lines with UWB antennas*

Long delay lines must be integrated with UWB antennas to increase the number of possible IDs, i.e. the number of structural tag mode possible delays (since the resolution is finite). This is not an easy process, since when integrating the line the original dimensions of the antenna and its ground plane are modified, affecting its performance. Next, a number of examples with antennas discussed section 2.5.1 will be considered to integrate long delays. Some of these will be fabricated on Rogers RO4003C substrate (see Table 2.3) to verify the simulations.

The first design is based on the broadband eccentric annular monopole UWB antenna [SHU 01]. It is shown in Figure 2.21. The ground plane has significant effects on planar UWB monopole antenna properties. This is due to the fact that the ground plane introduces extra resonant modes, changes the current distribution and hence distorts the antenna performance. A negative impact on antenna performance is also obtained when it is connected to long delay lines with meander shapes. These effects are shown in the simulated return loss of the tag in Figure 2.22(a). Simulations are done with Agilent Momentum Electromagnetic Simulator. Another example is shown in

Figure 2.22(b), with a narrower meander line. In this case, the effect is more noticeable. It can be observed that when the UWB antenna is fed with a thru line, the effect of the slots is not crucial. On the contrary, when there is a long meander-shaped delay line connected to the antenna, the slots become essential. To solve these problems two slots have been introduced between the antenna and the ground plane (see dashed circles in the inset of Figure 2.21), similar to what was done in [LU 11, LU 10]. When the slots are present, the delay line has a much smaller influence on the antenna performance. This permits us to integrate large ground planes with monopole UWB antennas. The main advantage is that long delay lines with an arbitrary shape can be connected to the antenna and, as a consequence, it permits us to increase the number of bits that can be coded. This is the main difference of this tag compared with scattering antennas presented in the literature, which integrate short delays [HU 10] or the delay is synthesized by means of a coaxial cable connected to the UWB antenna [LAZ 11, RAM 11]. The tag size is 10 cm × 10.8 cm and its simulated and measured $|S_{11}|$ is shown in Figure 2.23(a). Figure 2.23(b) shows a schematic representation of the simulated current distribution of the tag design and without with slots (with 2.5 mm separation). As can be observed, the current is concentrated around the slots, minimizing the intensity in the transmission line, and thus minimizing the effect of the transmission line on the antenna.

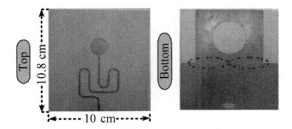

Figure 2.21. *Photographs of the manufactured tag based on a broadband eccentric annular monopole antenna with ground plane slots. Dimensions are in millimeters. For a color version of the figure, see www.iste.co.uk/ramos/rfid.zip*

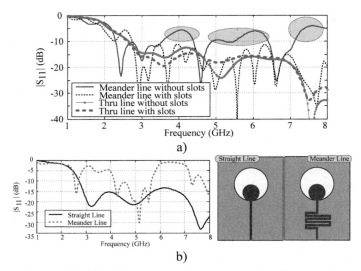

Figure 2.22. *Simulated return loss to show the effects of the slot that separates the antenna and the delay line with a) a wide meander line and b) a narrow meander line. For a color version of the figure, see www.iste.co.uk/ramos/rfid.zip*

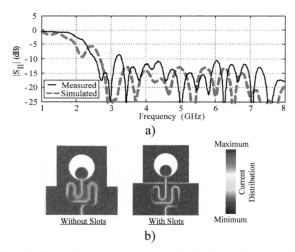

Figure 2.23. *a) Simulated and measured |S₁₁| for the tag based on a broadband eccentric annular monopole with a meandered line with slots; b) schematic representation of the current distribution for the tag with a meandered line without and with slots. For a color version of the figure, see www.iste.co.uk/ramos/rfid.zip*

Based on the same broadband eccentric annular monopole, another design is considered without slots in the ground plane. The tag is based on a square-shaped transmission line that rounds the monopole. The antenna size has been increased to slightly reduce its center frequency. This design takes advantage of the large ground plane on the back of the antenna. Figure 2.24 shows the tag layout and Table 2.4 shows its dimensions. Figure 2.25 shows a photograph of the tag and Figure 2.26 compares the simulated and measured reflection coefficient ($|S_{11}|$) of the tag. The measured $|S_{11}|$ parameter slightly differs from the simulated result, but the overall response is well matched for the operation band.

Parameter	Dimensions (mm)	Parameter	Dimensions (mm)	Parameter	Dimensions (mm)	Parameter	Dimensions (mm)
L	100	L4	10.135	W2	70.7	W6	5.842
L1	58	L5	6.185	W3	10	W7	6.435
L2	42	L6	9.06	W4	90	Ri	8.95
L3	19.8	W1	10.475	W5	54.01	Ro	18.05

Table 2.4. *Dimensions of the square-shaped line tag in millimeters. Line width: 1.88 mm*

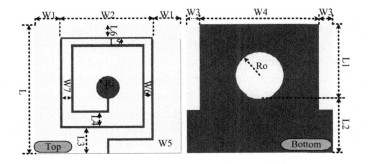

Figure 2.24. *Tag layout of the broadband eccentric annular monopole with the square-shaped line*

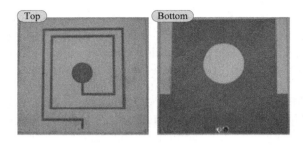

Figure 2.25. *Photograph of the square-shaped line tag. Tag size: 10 × 11 cm²*

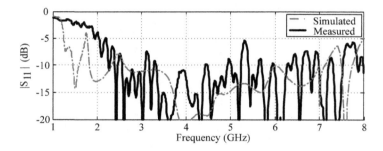

Figure 2.26. *Simulated and measured $|S_{11}|$ for the square-shaped line tag. For a color version of the figure, see www.iste.co.uk/ramos/rfid.zip*

Finally, Figures 2.27 and 2.28 show the case of a dipole and a Vivaldi UWB antenna, respectively. The dipole antenna is not heavily affected by the meander line. This is because the line is integrated in the same space of the access line, without modifying the geometry. However, the main problem in the dipole case is that the space is limited by the dimensions of the elliptic radiators, which in turn depend on the frequency of the antenna. The Vivaldi antenna, on the contrary, permits us to integrate a large ground plane without affecting the radiator. In addition, as shown in the layout of Figure 2.28(a), no slots are needed to separate the ground plane. The main disadvantage with this antenna, however, is that the radiation pattern is more selective toward angles near 90° in the H-plane (as shown in Figure 2.20(b)). This means that a tag based on a Vivaldi antenna would have difficulties in being read at certain tag-reader angles, but, on the other hand, would read better at its maximum radiation angle (due to the greater gain).

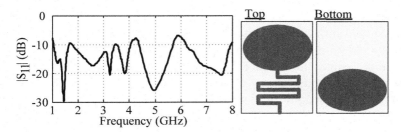

Figure 2.27. *Simulated* $|S_{11}|$ *and layout of a dipole-based tag with an integrated meander line*

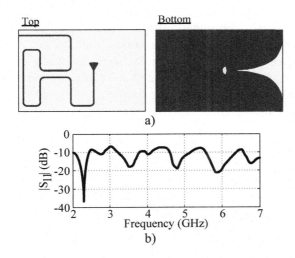

Figure 2.28. *a) Layout and simulated* $|S_{11}|$ *b) of a tag based on a Vivaldi antenna with an integrated meander line*

2.5.3. *Circularly polarized UWB RFID tags*

In order to be able to read an RFID tag in any orientation, a circular polarization is required in the antennas. This can be achieved by either reading the tag with circularly polarized UWB antennas in the reader, or by creating circularly polarized UWB tags.

Although circularly polarized narrowband antennas are well-known and studied, with UWB antennas it is a more complex process due to the large bandwidth required. An axial ratio below 3 dB is required over a >1 GHz bandwidth. Also, as they will be RFID tags, they should be small antennas (the antenna with the integrated delay line should not be bigger than a credit card). Several works have tried different designs to enhance the axial ratio, such as embedding two inverted-L grounded strips around two opposite corners of the slot [POU 11], embedding a T-shaped grounded metallic strip [XU 07], embedding a spiral slot in the ground plane [MAS 13] or embedding a lightning-shaped feed line [SZE 10].

Two examples are shown in the following. Both examples are optimized with Agilent Momentum Electromagnetic Simulator and fabricated Rogers RO4003C (see Table 2.3). The first antenna consists of two inverted-L grounded strips, which provide the circular polarization. The layout with the main dimensions is shown in Figure 2.29(a). The parameters have been optimized to center the operating frequency between 2.5 and 5.5 GHz (within the Novelda and Time Domain radars operating frequency). Figure 2.29(b) shows the simulated $|S_{11}|$ parameter and the simulated axial ratio.

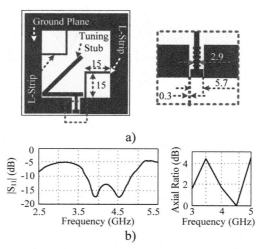

Figure 2.29. *a) Layout of the circularly polarized UWB antenna based on two inverted-L grounded strips (dimensions in millimeters; b) simulated $|S_{11}|$ and axial ratio*

A coplanar waveguide (CPW) transmission line is integrated with the shown antenna. The CPW line width is calculated to match the impedance of the antenna (21.47 Ω). The transmission line is added bearing in mind the slots which separate the antenna and the line, introduced with the broadband eccentric annular monopole tag with a meandered line (section 2.5.2). Figure 2.30(a) shows the layout and photograph of the tag. The slots which separate the ground plane are pointed out with blue-dashed circles. Figure 2.30(b) shows the simulated $|S_{11}|$ and axial ratio. As can be observed, the performance has been slightly reduced from the original design of the antenna, but still works over the band of interest. The main advantage of this tag, besides its circular polarization, is that it is manufactured on a single layer. In large-scale manufacturing, with processes such as inkjet printing (see section 2.5), this permits a simpler, lower cost fabrication.

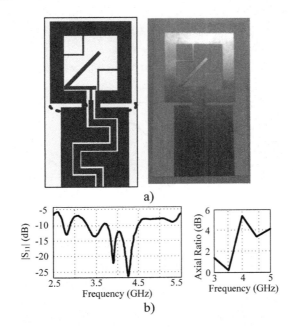

Figure 2.30. *a) Layout and photograph of the circularly polarized UWB tag based on two inverted-L grounded strips; b) simulated $|S_{11}|$ and axial ratio*

The second example uses a broadband eccentric annular monopole antenna with an L-shaped excitation. The layout of the tag and its dimensions, along with photographs of the top and bottom faces, are shown in Figure 2.31(a). Figure 2.31(b) shows the simulated $|S_{11}|$ parameter and the simulated axial ratio. As can be observed, there is a slightly better performance than the previous CPW-fed antenna. However, the tag size in this case (about $182 \times 124 \text{ mm}^2$) is unpractical. This tag is therefore used as a reference tag with circular polarization.

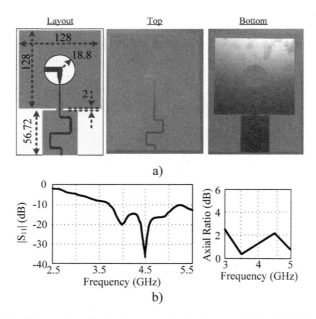

Figure 2.31. *a) Layout and photograph of circularly-polarized UWB tag based on a broadband eccentric annular monopole with L-shaped excitation (dimensions in millimeters; b) simulated $|S_{11}|$ and axial ratio*

2.6. Characterization of chipless time-coded UWB RFID tags

The chipless time-coded UWB tags discussed in section 2.5 are characterized in the time domain. The resolution to code an ID, the influence of the material where the tag is attached to, the tag-reader angle and the polarization are studied. Finally, some tags are realized in a flexible substrate in order to study the effects of bending.

2.6.1. *Time-domain response: distance and resolution*

Four tags based on the broadband eccentric annular monopole with the meandered line and slots (A, B, C and D) are fabricated. They have identical size with different delay line lengths L (209, 237, 265 and 293 mm, respectively). The delay lines are terminated with an open circuit. Figure 2.32 shows the four prototypes. Figure 2.33(a) shows the measured tag responses using the step-frequency approach, and Figure 2.33(b) shows the same responses using the impulse approach (with the Geozondas radar, see section 2.3.2). In both measurements, the CWT is applied, as described in section 2.4.2. The structural modes are identical for the four tags, since their shape and size and the tag-reader distance are the same. The tag modes, which depend on the transmission line length L, have a different delay with respect to their corresponding structural modes. The right inset in Figures 2.33(a) and (b) shows the zoomed tag modes.

Next, the tag based on the broadband eccentric annular monopole with the square-shaped line is measured using the Novelda radar. As shown in Figure 2.34, the fabricated tag is measured under three load conditions (states). As pointed out before, the tag mode depends on the load impedance, whereas the structural mode is independent of it. The first state is the tag without any element connected at the end of the delay line. The second and third states add, respectively, one and two coaxial line transitions at the end of the delay line. The three states are measured under the same conditions at a 50 cm reader-tag distance. Their measured responses are shown in Figure 2.35. It can be clearly seen that the tag modes (late peaks) are perfectly detected. They appear 6 ns after the structural mode (first peak). The tag modes corresponding to states 2 and 3 appear slightly after state 1 (the original). This is due to the delay added with the coaxial line transitions.

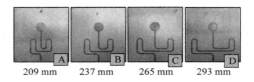

209 mm   237 mm   265 mm   293 mm

Figure 2.32. *Photographs of the four prototypes (a–d) of time-coded chipless tags based on broadband eccentric annular monopoles with a meandered line*

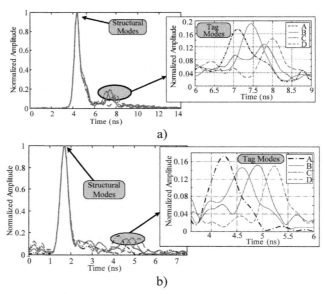

Figure 2.33. *Processed time-domain measurement of the tags based on broadband eccentric annular monopoles with a meandered line a) using the step-frequency approach and b) the impulse-based approach. For a color version of the figure, see www.iste.co.uk/ramos/rfid.zip*

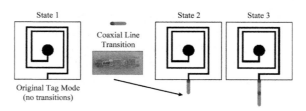

Figure 2.34. *Scheme of the emulated states for the square-shaped line tag. For a color version of the figure, see www.iste.co.uk/ramos/rfid.zip*

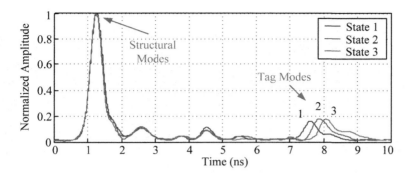

Figure 2.35. *Measured time-domain responses of the three tag states at a distance of 50 cm. For a color version of the figure, see www.iste.co.uk/ramos/rfid.zip*

Figure 2.36 shows the measured time-domain response for distances from 50 to 130 cm, in steps of 10 cm. The early reflections (marked with green arrows) correspond to the structural mode, while the late reflections (marked with red arrows in the top right inset) correspond to the tag modes. Due to the freespace propagation attenuation, the structural and tag modes decrease when the distance is increased. Amplitudes are normalized with respect to the 50 cm structural mode, which is the maximum amplitude obtained since it is at the closest distance. Both modes are detectable for all distances.

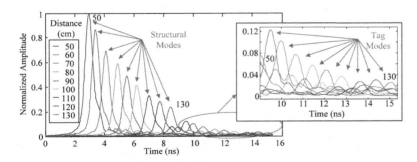

Figure 2.36. *Measured time-domain response of the square shaped line tag for distances from 50 to 130 cm. For a color version of the figure, see www.iste.co.uk/ramos/rfid.zip*

The distance is then increased until the tag mode cannot be distinguished to check the maximum read range. Figure 2.37 shows the time-domain response at a distance of 1.8 m, normalized with respect to the 50 cm structural mode amplitude. The time window has been shifted in order to measure the tag mode at 1.8 m, since the tag mode for 1.3 m appeared at the very end of the window used, as shown in Figure 2.36. In this last case, the tag mode can still be distinguished, but the amplitude is much lower than the ones from Figure 2.36. This is the limit for the presented system, which is a very large read range for a chipless integrated tag. Previous work [LAZ 11] shows that an emulated tag (using a UWB antenna connected to a coaxial transmission line) can be read up to 2 m. However, when measuring a real tag where the delay is integrated into the substrate the response worsens. Moreover, the integrated delay line on Rogers RO4003C has more losses than a coaxial transmission line, reducing the amplitude of the tag mode. Nevertheless, the distance of 1.8 m is much higher than the distances achieved for frequency-coded integrated chipless tags [PRE 09].

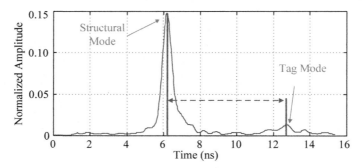

Figure 2.37. *Measured time-domain response of the square-shaped line tag at a distance of 180 cm. For a color version of the figure, see www.iste.co.uk/ramos/rfid.zip*

In order to measure the resolution experimentally, three tags based on the broadband eccentric annular monopole with the meandered line and slots, with a simulated delay difference of 50 ps between them, are fabricated. Figure 2.38 shows their measured responses (measured with the step-frequency technique shown in Figure 2.38(a) and with the impulse technique shown in Figure 2.38(b)). The right insets show the zoomed tag modes. Table 2.5 depicts the measured time differences between structural and tag modes. It can be clearly seen that a 50 ps delay cannot be distinguished and a 100 ps delay can be distinguished with a measurement error under 13%. Taking this result into account and since tag sizes should be as small as possible, it is obvious that a large number of states are not feasible. It is also important to note that the time resolution is inversely related to the bandwidth of the system. For a 7.5 GHz bandwidth (from 3.1 to 10.6 GHz), a resolution of about 66 ps is expected. Hence, this demonstrates that, although long delay lines have been integrated in time-coded chipless RFID tags, this topology might be more suitable for wireless sensor applications rather than for traditional item tagging. Frequency-coded tags, on the contrary, have an ID space much larger [PRE 09, VEN 11].

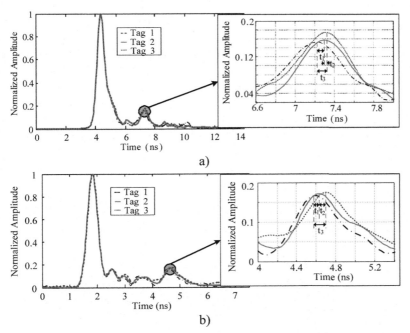

Figure 2.38. *Measured time responses of three tags with a simulated delay difference of 50 ps between them to evaluate time resolution. Measured with the step-frequency technique a) and the impulse technique b). For a color version of the figure, see www.iste.co.uk/ramos/rfid.zip*

Time delay (ps)	Theoretical	Step-frequency approach	Impulse-based approach
t_1	50	77.76	38.46
t_2	50	15.55	73.26
t_3	100	93.31	112.7

Table 2.5. *Theoretical and measured time delays for three tags with a simulated delay difference of 50 ps to evaluate time resolution*

2.6.2. *Angular behavior*

An RFID tag can be oriented in any angle when it is read, and a null reading angle can affect the read range of the system. Therefore, the effects of the read angle are studied in the following. To obtain a long read range, the tag mode amplitude should be as large as possible. This can be achieved by reducing the dimensions of the tag, and specially reducing the metallic area. Three tags are to be compared in the following. They are shown in Figure 2.39: the square-shaped line tag from section 2.5.2, a miniaturized monopole tag (addressed as "small tag") and a tag based on a Vivaldi antenna. Figure 2.40 shows the time-domain response for the three tags at 0°. They are read using the Novelda radar described in section 2.3.2. The ratio between the structural and tag modes is 18% for the square-shaped line tag, 50% for the small tag and 294% for the tag based on a Vivaldi antenna. As expected from its RCS, the structural mode of a Vivaldi antenna is very small and this results in a tag mode larger than the structural mode.

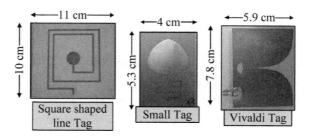

Figure 2.39. *Tags used to characterize the angular behavior*

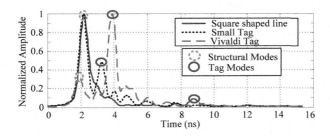

Figure 2.40. *Time-domain response for the square-shaped line, the small tag and the Vivaldi tag. For a color version of the figure, see www.iste.co.uk/ramos/rfid.zip*

Figure 2.41 shows the amplitude of the structural and tag modes as a function of the tag-reader angle in H-plane. It has been obtained from the time response for each angle, such as the 0° angle shown in Figure 2.40. Here, the structural mode corresponds to the RCS of the tag and the tag mode corresponds to the radiation pattern of the tag antenna. Figure 2.42 shows the ratio between the tag and structural modes, also as a function of the tag-reader angle in the H-plane. It compares the ratio between modes for the three tags presented in this section. As can be observed, the small tag maintains a constant ratio for all orientations. The Vivaldi tag, on the contrary, shows a very large ratio for its maximum radiation angle. Finally, the square-shaped delay line tag shows a smaller ratio than the other cases due to its larger size.

2.6.3. *Influence of materials*

RFID tags are intended to be attached to materials. Depending on the material, the tag performance can be affected. This effect is well known in UHF RFID [LOR 11], where UHF antennas are detuned. It will be studied in the following for time-coded chipless tags, where UWB antennas are used instead of narrowband antennas. The tag used to study the influence of materials is based on the tag proposed in [HU 08]. In this case, the profile of the monopole ground plane has been modified in order to improve its performance. As explained in section 2.5.2, the ground-plane shape of planar monopole antennas has significant effects on their properties, such as impedance bandwidth, radiation pattern, gain and time-domain response.

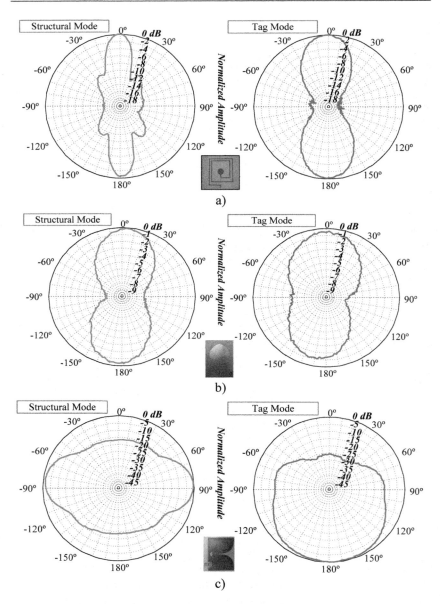

Figure 2.41. *Measured structural and tag modes for the square shaped line tag a) the small tag b) and the tag based on a Vivaldi antenna c) as a function of the H-plane angle. For a color version of the figure, see www.iste.co.uk/ramos/rfid.zip*

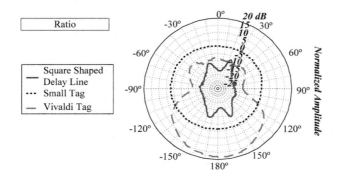

Figure 2.42. *Measured ratio between tag and structural modes for the three tags. For a color version of the figure, see www.iste.co.uk/ramos/rfid.zip*

Furthermore, the structural mode is usually much larger than the tag mode, and, at the end, this difference decreases the read range. Then, reducing this difference is a key aim in the design of time-coded chipless UWB tags. One way to achieve this is by reducing the structural mode. The tag proposed in [HU 08] decreases the structural mode by emptying the monopole circular radiating element and by reducing the tag size. The main disadvantage in reducing the tag size, however, is that a few number of IDs can be coded, since the transmission line length cannot be very large. The tag is fabricated on Rogers RO4003C substrate (see Table 2.3). Its simulated and measured reflect coefficient and photographs are shown in Figure 2.43.

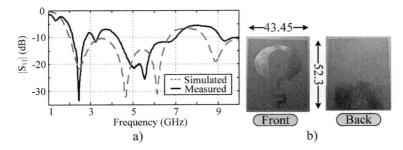

Figure 2.43. *a) Simulated (dashed red line) and measured (solid black line) $|S_{11}|$ parameter of the material characterization tag. b) Photograph and dimensions of the tag in millimeters. For a color version of the figure, see www.iste.co.uk/ramos/rfid.zip*

In order to study the influence of the material attached to the tag on the performance, several materials with the same size as the tag have been considered. This size is important since the structural mode also depends on the tag size and shape, and we want to study only the contribution due to the material. As shown in Figure 2.44, five typical materials to which RFID tags are attached have been chosen: cardboard, Teflon, Polyvinyl chloride (PVC) and two types of wood–particleboard (also known as chipboard, addressed in this work as *Wood1*) and strip wood (addressed here as *Wood2*).

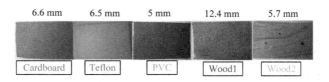

Figure 2.44. *Photograph of the attached materials and their thickness in millimeters. From left to right: cardboard, Teflon, PVC, particleboard wood (Wood1) and strip wood (Wood2) . For a color version of the figure, see www.iste.co.uk/ramos/rfid.zip*

Figure 2.45 shows the time-domain response of the tag in free space read from 40 to 200 cm in 20 cm steps, normalized with respect to structural mode at 40 cm. Both the structural (green arrow and circles) and tag (red arrow and circles) modes can be clearly detected with a 700 ps time difference between them and it is independent of the distance.

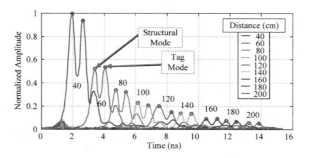

Figure 2.45. *Measured tag response for distances from 40 to 200 cm in 20 cm steps, without any material attached to it. For each distance, its corresponding structural (early, green) and tag (late, red) mode peaks are circled. The signals are normalized with respect to the 40 cm structural mode. For a color version of the figure, see www.iste.co.uk/ramos/rfid.zip*

Figure 2.46 shows the time-domain response at a fixed 40 cm distance, but in this case the tag is measured in free space and attached to the materials described in Figure 2.44. It should be noted that the tag must be attached to the material in its back side in order not to change the delay that identifies it (this phenomenon will be studied in detail in section 3.3). All the time-domain responses are normalized with respect to the free-space structural mode. It can be seen that the structural mode amplitudes increase for all cases except for the cardboard case. Since cardboard is a hollow material, its contribution to the structural mode is not noticeable. Therefore, the structural mode amplitude is very similar to the free space case. The thicker and the higher the permittivity of the material is, the higher the structural mode, as seen with the particleboard, Teflon and PVC. The tag mode, however, is barely affected by the attached material for all cases, except for the particleboard, which considerably reduces the tag mode amplitude. Particleboard is an engineered wood, which is made up of pressed wood particles and a synthetic resin. This synthetic resin clearly affects the radiation properties of a UWB antenna, reducing its performance.

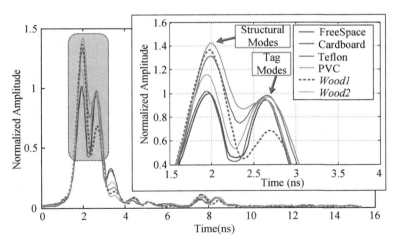

Figure 2.46. *Measured tag response at a fixed 40 cm distance when the tag is in free space and when it is attached to the materials shown in Figure 2.44. The signals are normalized with respect to the free space structural mode. The right inset shows the zoomed structural and tag modes. For a color version of the figure, see www.iste.co.uk/ramos/rfid.zip*

Next, the tag is read from 40 to 200 cm in steps of 20 cm attaching the same materials and measuring the time-domain response for each material at each distance. Figure 2.47 shows the normalized amplitudes of the tag modes as a function of the tag-reader distance for all the materials and for the free space case. It can be observed that, again, the tag mode is barely affected by the attached material for all cases except for the particleboard (*Wood1*) due to the losses of the synthetic resin, which is used to manufacture particleboard wood.

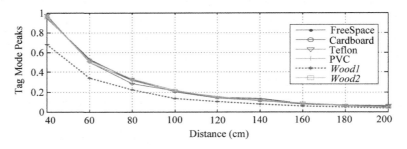

Figure 2.47. *Tag mode peak amplitudes for distances from 40 to 200 cm in 20 cm steps depending on the material attached to the tag. For a color version of the figure, see www.iste.co.uk/ramos/rfid.zip*

Figure 2.48 shows the normalized amplitudes of the structural modes as a function of the tag-reader distance for all the materials and for the free space case. A similar trend as that observed in Figure 2.46 for a fixed distance case can be seen for all distances: the cardboard and free space cases present similar structural modes, whereas the other materials increase it.

Figure 2.49 shows the ratio between structural (Figure 2.48) and tag (Figure 2.47) modes. Ideally, the ratio should be as great as possible in order to maximize the read range. Also, this ratio should be constant for any given distance, meaning that the read range of the system depends on the transmitted power rather than on the tag backscattering performance. A linear regression has been calculated for each material and is also shown. The linear regression parameters (slope and offset) are given in detail in Table 2.6. The goal is a zero slope (meaning the ratio is constant for all distances) and an offset as great as possible (meaning a high ratio value). It can be seen that attaching a material to the tag slightly worsens its performance due to the increase in the structural mode. However, when the tag is attached to particleboard, the performance is significantly reduced (it has an offset of 0.4510 with respect to the 0.9361 in the free-space case). This reduction is due to both the increase in the structural mode (thicker tag) and the decrease in the tag mode (due to the losses of synthetic resin).

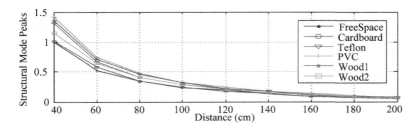

Figure 2.48. *Structural mode peak amplitudes for distances from 40 to 200 cm in 20 cm steps depending on the material attached to the tag. For a color version of the figure, see www.iste.co.uk/ramos/rfid.zip*

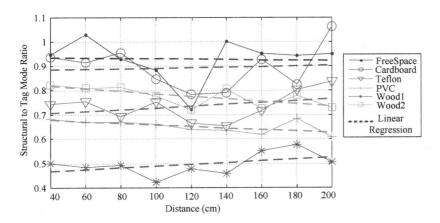

Figure 2.49. *Structural-to-tag mode ratios for distances from 40 to 200 cm in 20 cm steps depending on the material attached to the tag (solid lines). Linear regression for each material (dashed lines). For a color version of the figure, see www.iste.co.uk/ramos/rfid.zip*

Material	Structural-to-tag mode ratio linear regression parameters	
	Slope (cm^{-1})	Offset
Free space	-6.5087×10^{-5}	0.9361
Cardboard	1.1369×10^{-4}	0.8795
Teflon	3.7808×10^{-4}	0.6897
PVC	-3.1457×10^{-4}	0.6895
Particleboard wood (*Wood1*)	3.7704×10^{-4}	0.4510
Strip wood (*Wood2*)	-5.1088×10^{-4}	0.8367

Table 2.6. *Linear regression parameters for the structural to tag mode ratio as a function of the distance and attached material*

Figures 2.50 and 2.51 show the structural and tag modes as a function of the tag-reader H-plane angle (φ), normalized with respect to the maximum structural and tag modes in free space, respectively. This is shown for the free space case and the particleboard (*Wood1*) case. The structural mode in Figure 2.50 can be seen as the RCS of the tag. It is increased due to the presence of the particleboard, especially in 90°/–90° orientations, where it is very low for the free space case. As a first approximation, the structural mode is similar to a flat plate. Thus, its RCS is given by $RCS = 4\pi(A\cos(\varphi))^2 / \lambda^2$, where A is the area of the plate and $\lambda = c / \left(f\sqrt{\varepsilon_r} \right)$ is the wavelength (c is the speed of light in vacuum, f is the operating frequency and ε_r is the dielectric permittivity). The wavelength λ decreases when ε_r increases, and therefore the RCS is larger with the attached material. In addition, the shape of structural mode diagram changes because of the constructive interference between the front and back reflections due to the thickness of the material. The tag mode in Figure 2.51 can be seen as the radiation pattern of the antenna. In this case, the pattern changes from an omnidirectional pattern to a more directive pattern.

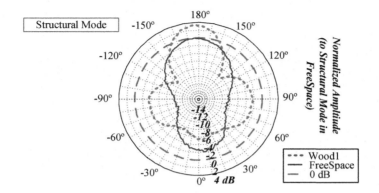

Figure 2.50. *Structural mode as a function of the tag-reader H-plane angle depending on the material used. All amplitudes are normalized with respect to the maximum of the structural mode in free space. For a color version of the figure, see www.iste.co.uk/ramos/rfid.zip*

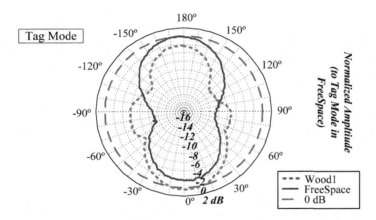

Figure 2.51. *Tag mode as a function of the tag-reader H-plane angle depending on the material used. All amplitudes are normalized with respect to the maximum of the tag mode in free space. For a color version of the figure, see www.iste.co.uk/ramos/rfid.zip*

2.6.4. *Polarization*

The polarization of several tags will be characterized in the following. A small monopole (section 2.6.2), a Vivaldi tag (section 2.5.2) and the circularly-polarized tags (section 2.5.3) are compared. Figure 2.52(a) shows a scheme of the measurement: the tags are placed on a rotatory table, oriented toward the reader's antennas on their maximum H-plane angle. Figure 2.52(b) shows the tag mode as a function of the polarization angle. As can be observed, the Vivaldi antenna is strongly polarized, and the tag mode cannot be read for cross-polarization angles (−90°/90°). The small monopole also shows that for cross-polarization angles (0°/180°) there is a significant reduction in the tag mode amplitude. The tag mode amplitudes of the circularlypolarized tags show a fairly constant behavior for all polarizations.

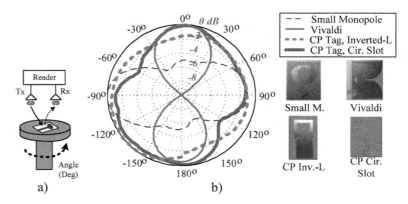

Figure 2.52. *a) Scheme of the polarization measurement; b) angular measurement of the tag mode as a function of the polarization for a small monopole, a Vivaldi and two circularly-polarized tags. For a color version of the figure, see www.iste.co.uk/ramos/rfid.zip*

2.6.5. *Flexible substrates: bending*

One desirable feature with RFID tags is the ability to bend them and still obtain a reliable read. This is interesting when the tag is attached to non-flat surfaces, such as pipes, bottles or clothes. To this end, flexible substrates must be considered [RIA 09]. In UHF RFID, there is a shift toward lower frequencies in the tag operating frequency when it is bent [YAN 08]. Similarly as with UHF cases, this section will study the behavior of a chipless time-coded UWB RFID tag when it is bent.

Two tags are optimized with Agilent Momentum Simulator and fabricated on Rogers Ultralam 3850 substrate ($\varepsilon_r = 2.9$, tan $\delta = 0.0025$, substrate thickness of 100 µm and metallization thickness of 18 µm). This is a dual-layer flexible substrate based on a liquid crystalline polymer.

The first flexible tag consists of a microstrip broadband eccentric annular monopole antenna with a meandered line. A delay line of 1.5 ns has been added to separate the structural and tag modes. Two slots in the ground plane have also been inserted (see section 2.5.2). The tag layout, dimensions and photographs are shown in Figure 2.53(a), and its simulated $|S_{11}|$ parameter is shown in Figure 2.53(b). As can be observed, there is a good matching over the UWB radars operating band (3.1–5.6 GHz).

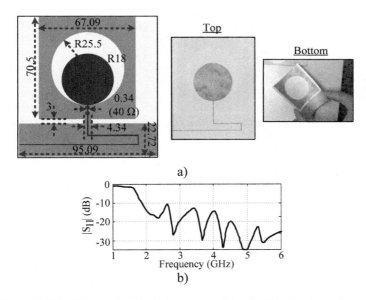

a)

b)

Figure 2.53. *a) Layout and photographs of the flexible tag based on a microstrip broadband eccentric annular monopole with a meandered line (the dimensions are in millimeters); b) simulated $|S_{11}|$ parameter*

The second tag consists of a CPW-fed antenna based on the design given by Tavassolian *et al.* [TAV 07]. In this case, the elliptical slot has been replaced by a circular slot. In addition, two slots have been added between the antenna and the ground plane of the transmission line, with a separation of 2 mm, as in section 2.5.2. The delay of the transmission line is 1 ns. The tag layout is shown in Figure 2.54(a). Figure 2.54(b) shows the simulated |S$_{11}$| parameter. As can be observed, the tag also works within the UWB radar frequency band.

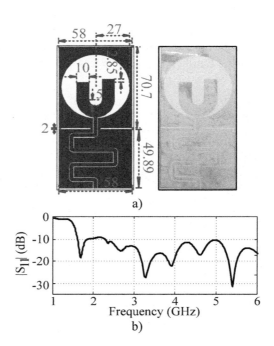

a)

b)

Figure 2.54. *a) Layout and photograph of the flexible tag based on a CPW-fed UWB antenna (dimensions in millimeters); b) simulated |S$_{11}$| parameter*

The tags are bent and measured with the Time Domain radar (see section 2.3.2) in a semi-anechoic environment. Figure 2.55 shows the measured time-domain response for the microstrip broadband eccentric annular monopole tag. As can be observed, the structural mode is greatly reduced, whereas the tag mode is barely affected when the tag is bent. The structural mode, as explained in section 2.6.1, depends on the RCS of the tag. Since the RCS area is noticeably smaller when the tag is bent, the structural mode is affected. The tag mode, on the contrary, is not shifted in time and its amplitude is not affected. Due to the use a large bandwidth, a detuning effect is not occurring as with narrowband UHF RFID tags.

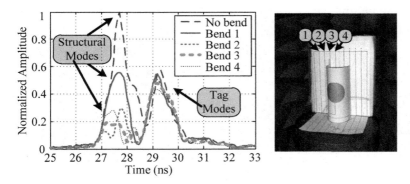

Figure 2.55. *Measured time-domain response of the flexible tag based on a broadband eccentric annular monopole as a function of the bending on the tag. For a color version of the figure, see www.iste.co.uk/ramos/rfid.zip*

Figure 2.56 shows the same measurement for the CPW-fed flexible tag. In this case, the effects of bending the tag clearly worsen the structural mode, generating a second reflection thatdoes not correspond to the tag mode. This effect is explained because the structural mode depends on the RCS of the shape of the tag, which changes from a flat shape to a semi-cylindrical shape. In the case of a cylinder, the waves are backscattered in all the directions, and not only toward the reader. As a consequence, the structural mode is lower than the flat case. The tag mode, however, shows the same behavior as with the broadband eccentric annular monopole case: it is barely affected by the bending, and not only the time delay but also the amplitude remains mostly the same.

Finally, a practical example will be studied with the broadband eccentric annular monopole flexible tag, since it has shown a better robustness in terms of structural mode, and a larger Structural-to-tag mode ratio. Figure 2.57 shows the measured time-domain response of the tag placed on top of a plastic (PET) bottle, when the bottle is empty and when it has been filled of water. As expected, the tag mode is greatly affected because of the water (as with many other RFID tags). However, since the structural mode remains, the reader could detect whether the bottle is empty/full by the presence or absence of the tag mode.

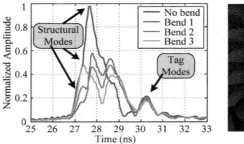

Figure 2.56. *Measured time-domain response of the flexible tag based on a CPW-fed UWB antenna as a function of the bending on the tag. For a color version of the figure, see www.iste.co.uk/ramos/rfid.zip*

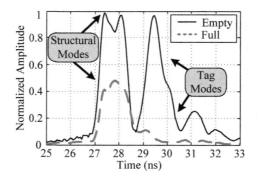

Figure 2.57. *Measured time-domain response of the flexible tag based on broadband eccentric annular monopole antenna bent on a plastic bottle, with and without water inside of it. For a color version of the figure, see www.iste.co.uk/ramos/rfid.zip*

2.7. Conclusions

This chapter has presented the study of chipless time-coded UWB tags. A circuit theory has been developed in order to understand the signals occurring between the reader and the tag. Two approaches (in frequency and time) for realizing readers have been presented, comparing the advantages and disadvantages between them. It has been demonstrated that potentially low-cost readers can be realized by using commercial UWB-pulsed radars. To achieve this, several signal processing techniques have been presented. The design of chipless time-coded UWB RFID tags has also been addressed, specifically by integrating long delay lines with UWB antennas. Finally, the tags have been characterizedby obtaining the following results:

– the number of possible IDs is very limited (few bits) compared with frequency-coded tags;

– the read range of time-coded tags is much larger: up to several meters;

– depending on the antenna structure, the angular behavior changes. A tag that can be read at any tag-reader angle will have a smaller read range;

– chipless time-coded UWB tags are barely affected by the material they are attached to. This is a noticeable advantage over narrowband RFID tags, which are detuned by the material;

– it is possible to achieve circular polarization with time-coded UWB RFID tags by integrating delay lines with circularly-polarized UWB antennas. However, the tag size increases noticeably, and the antenna bandwidth is smaller;

– when using a flexible substrate to manufacture chipless time-coded tags, there is a possibility to bend the tags on round objects. The tag modes are neither shifted nor are their amplitudes diminished. However, extreme bending angles do affect the structural mode amplitude and time delay.

3

Wireless Sensors Using Chipless Time-coded UWB RFID

3.1. Introduction

Completely passive battery-free wireless sensors are desirable in remote sensing applications where long-term controlling and monitoring of the environment take place. In addition, since they require neither wiring nor batteries, they can be used in hazardous environments, such as contaminated areas, under concrete, in chemical or vacuum process chambers and also in applications with moving or rotating parts.

Chipless RFID sensor tags could be an interesting alternative for passive wireless sensing. They consist of integrating a passive sensor into a chipless tag. The integration of sensors in chipless tags as well as the development of their reading systems could result in a challenging and unique opportunity in several applications such as the ones addressed above. For instance, an identification sensor system platform at 915 MHz for passive chipless RFID sensor tags is proposed in [SHR 09] to sense ethylene gas concentration. A wireless temperature transducer based on micro-bimorph cantilevers and split ring resonators at 30 GHz is presented in [THA 10]. Passive wireless pressure micromachined sensors are proposed in [JAT 09] and [FON 02]. A 13-bit frequency-coded chipless sensor based on silicone nanowires to detect both temperature and humidity is presented in [VEN 12]. Chipless RFID sensor tags where the identification code generation is carried out using SAW devices have also been addressed in a number of works; three examples of temperature sensors based on SAW technology are proposed in [REI 04], [DOW 09] and [ALO 10].

This chapter considers chipless time-coded UWB RFID tags for wireless sensing. Two main strategies are studied: amplitude-based and delay-based. These strategies rely, respectively, on changing the tag mode amplitude or time delay depending on a physical parameter. Two sensor examples of each strategy are presented: temperature sensors and permittivity (for concrete composition detection) sensors. The chapter is organized as follows:

– section 3.2 presents amplitude-based chipless time-coded UWB sensors, specifically the following:

- a temperature sensor using off-the-shelf positive temperature resistors;

- temperature threshold detectors using shape memory alloys and using a thermal switch,

- detection techniques for threshold detectors,

- self-calibration techniques for chipless amplitude-based sensors;

– section 3.3 presents delay-based chipless time-coded UWB sensors, specifically, a permittivity sensor that enables concrete composition detection;

– finally, section 3.4 concludes the chapter.

3.2. Amplitude-based chipless time-coded sensors

3.2.1. *Principle of operation*

In section 2.2 the tag mode amplitude was said to depend on the reflection coefficient connected at the end of the transmission line. Figure 3.1(a) shows the circuit scheme where the tag mode amplitude is changed as a function of this reflection coefficient. The amplitude variations and the possibility of modulating them by means of a resistive sensor are shown in Figure 3.1(b). In this figure, several surface-mount (SMD, 0603) resistors are soldered at the end of the transmission line (Z_{LOAD}), and the corresponding time-domain signal is obtained. These measurements are carried out at a sensor-reader distance of 50 cm, with the frequency-step approach (VNA, section 2.3.1) and the broadband eccentric annular monopole antenna with a meandered-line slot (see section 2.5.2). One sensor can be identified from the other by using the delay between the structural and tag modes.

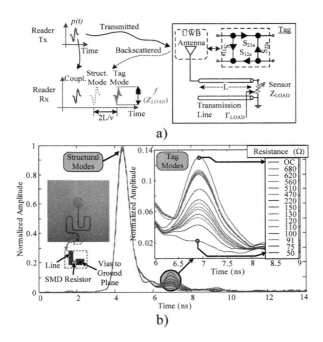

a)

b)

Figure 3.1. *a) Circuit scheme for an amplitude-based chipless RFID sensor tag. b) Measured amplitude variations of the tag mode when soldering several SMD resistors at the end of the transmission line (resistance values from 50 to 680 Ω, and an open circuit, OC). For a color version of the figure, see www.iste.co.uk/ramos/rfid.zip*

The structural modes remain identical for all resistors. All the tag modes appear at the same time (the transmission line length L is constant) and their amplitudes depend on the value of the resistor. It can also be observed that the higher the resistance is with respect to the transmission line characteristic impedance ($Z_c = 50$ Ω), the lower the reflection coefficient variation for a given temperature change. This saturation effect is expected since the tag mode amplitude depends directly on the reflection coefficient Γ_{LOAD}.

The resistance value can be obtained from the measured ratio between the structural and tag modes, ST_{RATIO}. This ratio is obtained from the peak of the tag mode amplitude normalized with respect to the peak of the structural mode amplitude. Then, a calibration technique similar to time-domain reflectometry (TDR) calibration is performed. The ST_{RATIO} for a given load resistance R can be expressed as:

$$\frac{ST_{\text{RATIO}}(R)}{ST_{\text{RATIO}}(R_{\text{MAX}})} = \frac{\Gamma_{\text{LOAD}}(R)}{\Gamma_{DC}(R_{\text{MAX}})},$$
[3.1]

where R_{MAX} is an arbitrary known resistance and $\Gamma_{DC}(R_{\text{MAX}})$ is the ideal (DC) reflection coefficient at R_{MAX}. In order to minimize measurement errors, a value of R_{MAX} that leads to a value of Γ_{DC} close to unity (open-circuit (OC) case) is chosen. In practice, a known high-value resistor can be used (here $R_{\text{MAX}} = 680 \ \Omega$, as shown in Figure 3.1(b)). Γ_{LOAD} are the measured reflection coefficients associated with each load resistance (R), $\Gamma_{\text{LOAD}} = (R - Z_c)/(R + Z_c)$ and are obtained from [3.1]. Using the tag to structural mode ratios, the measurement of the reflection coefficient (and thus the calibration) is theoretically independent from the tag-to-reader distance, as it will be addressed in section 3.2.2. Expression [3.1] predicts a linear model between ST_{RATIO} for a given resistance and its load reflection coefficient. Finally, the estimated resistance is calculated as:

$$R_{\text{ESTIMATED}} = 50 \times \frac{\left(1 + \Gamma_{\text{ESTIMATED}}\right)}{\left(1 - \Gamma_{\text{ESTIMATED}}\right)},$$
[3.2]

where $\Gamma_{\text{ESTIMATED}}$ is the load reflection coefficient computed from the linear regression of Γ_{LOAD} as a function of Γ_{DC}. In this way, measurement errors are reduced. Figure 3.2 shows the estimated resistance $R_{\text{ESTIMATED}}$ as a function of the real (soldered) resistance. This result demonstrates the feasibility of integrating resistive amplitude-based sensors in time-coded chipless tags to remotely detect resistance changes.

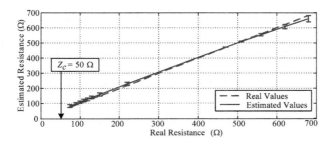

Figure 3.2. *Measured estimated resistance as a function of the real soldered resistance. For a color version of the figure, see www.iste.co.uk/ramos/rfid.zip*

3.2.2. *Temperature sensor based on chipless time-coded UWB tags*

In this case, a platinum positive temperature sensor (PTS) from Vishay Beyschlag is used as the resistive load. There are several resistance ranges: 100, 500 and 1000 Ω [VIS 14]. As shown in Figure 3.1, the 100 Ω sensor (addressed as PTS100, 0603) is the most suitable since its sensitivity to detect variations in the tag mode amplitude is larger when $Z_c = 50\ \Omega$. Figure 3.3 shows the measured values from the manufacturer's datasheet on top of the expected values from expression [3.3] that relates the resistance and temperature (where $A = 3.9083 \times 10^{-3}\ C^{-1}$, $B = -5.775 \times 10^{-7}\ C^{-2}$ and $R_0 = 100\ \Omega$ for the PTS100):

$$R_T = R_0\left(1 + AT + BT^2\right). \hspace{2cm} [3.3]$$

Since there is a very small error (below 5 mΩ for a 50 Ω range) between both, the equation can be used to obtain the equivalent temperature values from the measured resistance values.

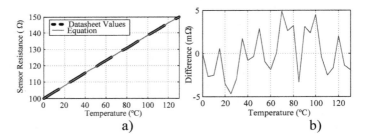

Figure 3.3. *a) Comparison of the datasheet values and the equation for the relation between the PTS100 resistance and temperature. b) Absolute error between both curves*

3.2.2.1. *Sensor design and characterization*

To characterize the PTS at a high frequency, it is connected as a load of a transmission line and is measured from 0 to 4 GHz. A custom calibration kit is designed (see Figure 3.4) to perform an open-short-load (OSL) calibration [AGI 04]. The temperature is then increased with a heatgun from 30 to 130°C in 5°C steps and the reflection coefficient is measured at each step. Figure 3.5 shows the measured reflection coefficient as a function of frequency for each temperature. The variation (and thus the sensitivity) decreases when the frequency increases. However, for the 1–3.5 GHz range, the variation is clearly detected. A tag based on the broadband eccentric annular monopole antenna connected to a meandered line including separation slots (see section 2.5.2) is designed. Its dimensions are scaled in order to decrease its operation frequency. The new sensor tag dimensions are 10×13.65 cm^2 and it is shown in Figure 3.6.

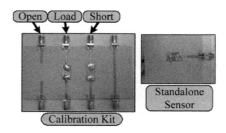

Figure 3.4. *Calibration kit used to measure the sensor reflection coefficient*

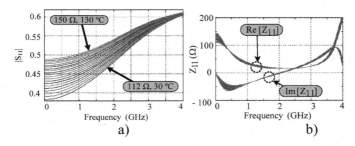

Figure 3.5. a) Measured variation in the Vishay PTS 0603 (100 Ω) sensor reflection coefficient as a function of frequency and temperature and b) corresponding sensor impedance. For a color version of the figure, see www.iste.co.uk/ramos/rfid.zip

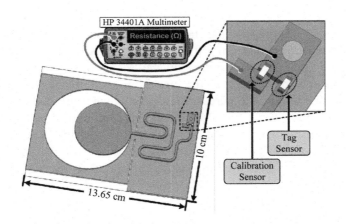

Figure 3.6. 3D image of the tag sensor and schematic of the calibration set-up

3.2.2.2. Results

The sensor tag is measured in a typical environment (laboratory). All measurements are done with the tag in mid-air, not attached to any material. The sensor tag is heated with an air flow coming from the heatgun up to the maximum temperature (130°C) and its response is measured while the tag cools down to room temperature. As shown in Figure 3.6, the Vishay PTS (100 Ω) sensor is soldered at the end of the transmission line. A second identical sensor is placed side-by-side for calibration purposes. This second sensor is connected to an HP-34401A multimeter that measures its resistance, thus giving the real resistance of both sensors as a function of the temperature. The thermal gradient between the two sensors is assumed to be very small (they are very close to each other) and thus, it is also assumed that the temperature is the same at the two sensors. Measurements are done simultaneously from the sensor tag with the UWB reader and from the multimeter. Of course, this second sensor is not needed in real applications since a calibration curve would be already available. Each measurement has its own sensor tag response signal (measured remotely using UWB backscattering) and its associated measured calibration resistance (obtained from the calibration sensor with the multimeter).

Figure 3.7(a) shows the tag mode amplitude variations when temperature changes from 30 to 130°C. These measurements are performed using a VNA (step-frequency technique, see section 2.3.1) and the signal processing techniques from section 2.4. As expected, the tag mode amplitude increases with the temperature. Here, the ideal reflection coefficient Γ_{DC} is calculated using the real resistance values obtained from the calibration sensor, instead of using known theoretical values (as performed with the discrete SMD resistors in section 3.2.1). Since the impedance of the PTS at the operation frequency differs from DC and it is not constant (see Figure 3.5), the slope of the ideal reflection coefficient Γ_{DC} is empirically adjusted to compensate that the amplitude of the reflected pulse is proportional to the average resistance over the antenna frequency band. In order to take this into account, an average reflection coefficient is empirically obtained as $\Gamma_{AVG}=F \times \Gamma_{DC}$, where $F = \Gamma_{DC}(R_{MAX})/\Gamma_{LOAD}(R_{MAX})$. Now $R_{MAX} = 150 \ \Omega$. The new estimated $\Gamma_{ESTIMATED}$ reflection coefficients and their corresponding estimated resistances $R_{ESTIMATED}$ can be calculated repeating the same procedure as in section 3.2.1. Figure 3.7(b) shows the ideal (Γ_{DC}), measured (Γ_{LOAD}) and estimated ($\Gamma_{ESTIMATED}$) reflection coefficients as a function of the real resistance.

Finally, the estimated temperature $T_{ESTIMATED}$ is obtained from $R_{ESTIMATED}$ using the equation given by the manufacturer and detailed in section 3.2.2. Figure 3.7(c) shows the estimated temperature as a function of the real temperature T_{REAL} (obtained from the calibration sensor) for a 50 cm tag-reader distance. The error bars at each point show the error with respect to the ideal case (when the estimated temperature equals the real temperature). An error below 0.6°C is obtained.

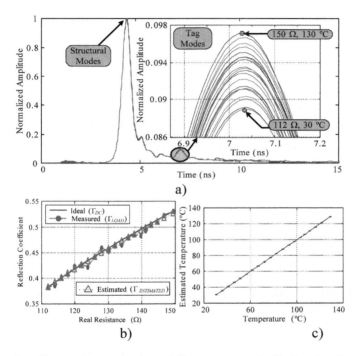

Figure 3.7. *Tag mode amplitude variations measured with the step-frequency technique a). Ideal, measured and estimated reflection coefficients as a function of the real resistance b). Estimated temperature as a function of the real temperature c). For a color version of the figure, see www.iste.co.uk/ ramos/rfid.zip*

Figure 3.8 shows the same results as shown in Figure 3.7, but now the measurements are carried out with the impulse technique based on UWB radar from Geozondas (see section 2.3.2). Since the impulse technique is nosier than the step-frequency technique, the error is higher. Now the error is below 3.5°C.

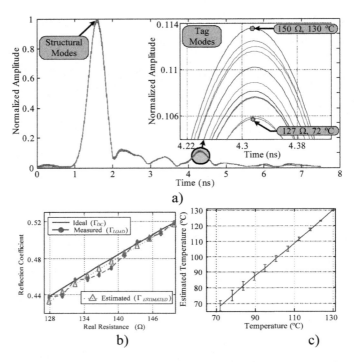

Figure 3.8. *Tag mode amplitude variations measured with the impulse-based technique a). Ideal, measured and estimated reflection coefficients as a function of the real resistance b). Estimated temperature as a function of the real temperature c). For a color version of the figure, see www.iste.co.uk/ ramos/rfid.zip*

Figure 3.9(a) shows the estimated temperature as a function of the real temperature for sensor-reader distances of 50, 65, 75 and 100 cm and using the step-frequency technique. The ideal curve corresponds to identical estimated and real sensor temperatures. In all cases, calibration has been performed using the data at 50 cm. It can be observed that the error in the temperature measurement increases when the tag-reader distance moves away from 50 cm. Figure 3.9(b) shows the same measurement, but now the calibration curve is obtained at each distance. It can be observed that the minimum error is at the last measured temperature, due to the calibration process explained above. It can also be noted that the error is considerably reduced when calibrating at each distance. Therefore, if we know the approximate distance between the reader and the sensor, it is better to use a previously stored calibration curve for that distance in order to minimize the error. If not, a calibration curve for a generic different distance can be used (see Figure 3.9(a)) at the risk of having a larger error. In a real environment, once the reader is installed, a measurement of the background (scene without the sensor) must be performed. Then, the sensor can be placed anywhere in the scene and temperature can be measured. The calibration parameters corresponding to the approximated expected reader-sensor distance must be used in order to minimize the measurement error.

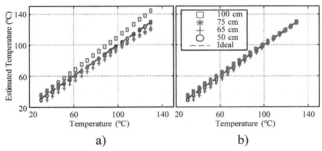

Figure 3.9. *Estimated sensor temperature as a function of the real temperature for sensor-reader distances of 50, 65, 75 and 100 cm using the calibration curve obtained at 50 cm a) and using the calibration curve obtained at each distance b). For a color version of the figure, see www.iste.co.uk/ ramos/rfid.zip*

Figure 3.10 shows the measured structural and tag modes as a function of the sensor-reader angle θ in H-plane in 2° steps. The sensor-reader distance is 50 cm. The transmission line has been loaded with an OC for this measurement, so the tag mode amplitude is at its maximum. Each mode is normalized to its own maximum. As introduced in Chapter 2, the structural mode corresponds to the radar cross-section (RCS) of the sensor tag and it depends on its shape, size and material. The tag mode corresponds to the radiation pattern of the sensor tag, since it depends on the radiation characteristics of the antenna.

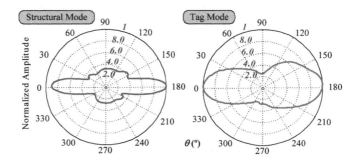

Figure 3.10. *Measured structural and tag modes as a function of the sensor-reader angle θ in H-plane in 2° steps. For a color version of the figure, see www.iste.co.uk/ramos/rfid.zip*

Figure 3.11 shows the ratio between the structural and tag modes from Figure 3.10 as a function of the sensor-reader angle in H-plane (in 2° steps). It is compared to the measured relative temperature error (RTE; measured in 15° steps). The RTE for a given angle is calculated from the mean relative error for each temperature at that angle: $RTE(\%)=100 \times mean\ [(T_{ESTIMATED} - T_{REAL})/T_{REAL}]$. It can be observed that the error increases when the ratio between the structural and tag modes decreases. The smallest errors appear for the highest ratios, at 30°, 150°, 210° and 330°. For a 90° and a 270° measurement angle, the error reaches up to 37%, which means that the temperature cannot be detected correctly. This is due to the minimum radiation angles of the sensor, as seen in the angular variation of the tag mode in Figure 3.10. However, for the remaining angles the error is below 10%, demonstrating that the sensor can be successfully read for most sensor-reader angles.

The sensing accuracy of the system is measured from a set of 300 measurements at a temperature of 28°C with a reader-sensor angle of 0°. Figure 3.12 shows the histograms obtained using the step-frequency technique and the impulse technique. An error below 0.4°C has been obtained for 97% of the measurements using the step-frequency technique and for 85% of the measurements using the impulse technique.

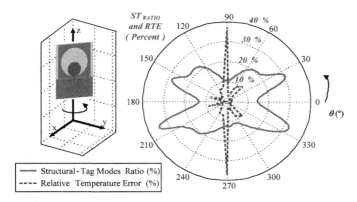

Figure 3.11. *Ratio between the measured structural and tag modes as a function of the sensor-reader angle in H-plane (in 2° steps) for a sensor-reader distance of 50 cm. It is compared to the measured relative temperature error (in 15° steps). For a color version of the figure, see www.iste.co.uk/ramos/rfid.zip*

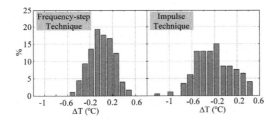

Figure 3.12. *Sensing accuracy of the system. Histograms obtained from 300 measurements at a temperature of 28°C with a reader-sensor angle of 0°*

The response time of the sensor tag is evaluated by cooling the sensor from 130 to 30°C. The sensor temperature is obtained as a function of the time. Figure 3.13 compares the response time as a function of the material attached to the tag. Four cases are considered: tag on still air, on a large metal plate, on a particleboard plate and on a PVC plate. Assuming a first-order model described by an exponential decay law similarly as done in [VIR 11], the response time of the sensor (t_r) can be defined as the time required to reach 95% of the final reading: $t_r(s/°C) = \Delta t_{95\%}/\Delta T = -(t_2 - t_1)/(T_2 - T_1)$, where t_1 and

t_2 are the instants when the temperature reaches $T_1 = 130°C$ and $T_2 = 130 - 0.95 (130 - 30) = 35°C$, respectively. The response times in sec/°C of the sensor are shown in the inset of Figure 3.13. These results show that the response time depends on the type of material where the sensor is attached (thermal conductor or insulator) and it is in the order of 1–2.6 sec/°C.

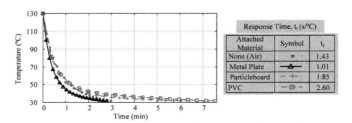

Figure 3.13. *Sensor temperature as a function of time when it is on air and attached to a large metal plate, to a particleboard plate and to a PVC plate. For a color version of the figure, see www.iste.co.uk/ramos/rfid.zip*

A comparison to other RFID temperature sensing circuits is shown in Table 3.1 in terms of technology, operation frequency, measured temperature range, sensing accuracy, response time and power consumption. Three main topologies are compared: temperature sensors based on SAW technology, based on RFID (LF and UHF) and, the presented here, based on chipless UWB RFID.

	Technology	Operation frequency (GHz)	Measured temperature range (°C)	Sensing accuracy (°C)	Response time	Measured power consumption
[REI 04]	SAW	2.422–2.45	+25 to +200	±0.2	N/A	-
[DOW 09]	SAW	2.4	+4 to +36	−0.8 to +0.6	3 min, from +4 to +36	-
[SAM 08]	UHF RFID	0.915	−20 to +50	2	21 s +48° to +20 23 s −26° to 35°	1.08 mW

[OPA 06]	LF RFID	100 −150 kHz	0 to + 100	2.5	N/A	N/A
[YIN 10]	UHF RFID 0.18 µm CMOS	0.915	−20 to +30	±0.8	N/A	1.5 µW
[ZIT 10]	0.35 µm CMOS	2.2	+20 to +100	0.3	N/A	34 µW
[VIR 11]	UHF RFID Alien Higgs 3	0.84–0.94	+33 to +58	±0.49	7 min to track 90% of range	N/A
[CHE 10]	UHF RFID 0. 18 µm CMOS	0.9	−40 to +100	±1.6	N/A	1.55–2.4 V and up to 260 µA
This book	Chipless UWB RFID	1–3.5	+25 to +130	<0.4 97% VNA 85% radar	3 min, from +130° to +30° (on air)	–

Table 3.1. *Comparison to other RFID temperature sensing circuits and systems*

3.2.3. *Temperature threshold detectors based on chipless time-coded UWB tags*

For some applications, an inexpensive temperature sensor, which informs whether a threshold has been surpassed, is enough. An alarm event can be triggered when the threshold has been surpassed. For instance, temperature threshold sensors using passive UHF RFID tags based on temperature sensitive shape memory polymers [BHA 10], printed nanostructures [GAO 10] and shape-memory-alloy (SMA) compounds [CAI 11] have already been proposed. Here, threshold temperature wireless sensors based on chipless time-coded UWB RFID are shown. Two sensors topologies are explained. The first sensor is based on a SMA, and the second on a commercial thermal switch. Finally, some detection techniques to improve the read range are explained, applicable due to the simplicity of these sensors.

3.2.3.1. *Threshold detector using shape memory alloys*

The first sensor consists of a time-coded chipless UWB RFID tag that is modified with a thermal actuator based on an SMA compound.

Shape-memory property is the ability of the material to undergo deformation at one temperature and then recover its original, undeformed shape upon heating above its transformation temperature. The SMA chosen in this work is Nickel Titanium (NiTi), also known as nitinol. It is a metal alloy of nickel and titanium [FU 04, HUA 10]. Nitinol is characterized by two solid-state phases, a martensitic phase and an austenitic phase, in which the structural and mechanical properties of the alloy greatly change. At high temperatures, nitinol assumes a cubic crystal structure referred to as austenite. At low temperatures, nitinol spontaneously transforms to a more complicated monoclinic crystal structure known as martensite. The temperature at which austenite transforms to martensite is generally referred to as the transformation temperature. Martensite's crystal structure (known as a monoclinic) has the unique ability to undergo limited deformation in some ways without breaking atomic bonds. When martensite is reverted to austenite by heating, the original austenitic structure is restored, regardless of whether the martensite phase was deformed. Thus, the nitinol is an SMA because the shape of the high-temperature austenite phase is remembered, even though the alloy is severely deformed at a lower temperature [HUA 10]. When the temperature threshold of the actuator is surpassed, the state of the tag is permanently changed.

Figure 3.14(a) shows a circuit scheme of the signals between the tag and reader, with the two tag topologies (addressed as options A and B) explained. In option A, the thermal switch short-circuits the delay line at its beginning. The tag mode cannot be detected since it is hidden under the (larger) structural mode. In option B, the delay line is load-matched at its end and no reflection is produced, so no tag mode is generated. When the thermal switch actuates, in both cases an open-ended delay line is connected to the UWB antenna and the tag mode can be detected. Figure 3.14(b) shows a photograph of the designed tag, and Figure 3.14(c) shows the two topologies in detail. Figure 3.15 shows the simulated $|S_{11}|$ parameter of the tag antenna. It consists of a small monopole (a variation of the ones presented in section 2.5.2), manufactured on Rogers RO4003C substrate (see Table 2.3). As can be observed, the resulting tag is well-matched from 2.75 to 7.5 GHz, which covers most of the frequency of interest.

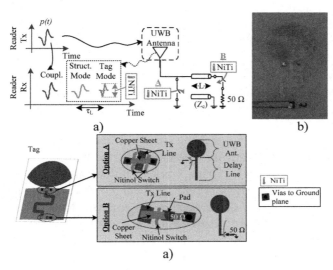

Figure 3.14. *a) Block diagram of the system. b) Photograph of the tag. c) Detail of the two options using the thermal switch made of nitinol as the actuator. Tag size: 26.6 mm × 38.2 mm. For a color version of the figure, see www.iste.co.uk/ramos/rfid.zip*

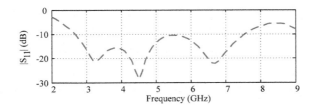

Figure 3.15. *Simulated |S₁₁| parameter of the tag*

The switch is based on a thermal actuator made of NiTi [FU 04] SMA wire connected to an adhesive conductive copper sheet. The switch uses the ability of this material to remember its original shape after being plastically deformed. A flow diagram of the alloy

[HUA 10] states is shown in Figure 3.16. The shape of the actuator is programmed at elevated temperatures above the transition temperature range (TTR). Then, the alloy is cooled below the TTR is plastically deformed to the switch initial position (copper sheet connected). When it is heated again above the transition temperature A_s (Austenite Start), it recovers the programmed shape and disconnects the copper sheet. Since the austenitic shape is stable, a successive fall of the temperature below the threshold will not modify the wire's shape anymore. The nitinol switch hence acts as a one-way actuator. Nitinol is typically composed of approximately 50–51% nickel by atomic percent (55–56% weight percent). Making small changes in the composition can change the transition temperature of the alloy significantly. We can control the transition temperature in nitinol to some extent, but convenient superelastic temperature ranges are from about −20°C to +60°C. In this work, a nitinol alloy from Kelloggs Research Labs with 55% weight percent is used, achieving an A_s temperature of 37°C (about body temperature). This alloy exhibits shape memory such that it is malleable at room temperature. It returns to shape just above body temperature. Thus, it is often used for dental products such as archwires and springs. The mechanical design of the actuator can be improved for commercial applications using advanced manufacturing techniques [FU 04].

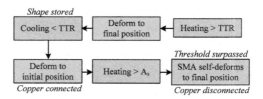

Figure 3.16. *Flow diagram of the NiTi states: programming of the shape, and tag operation to detect temperature*

As shown in Figure 3.17(a), when the temperature is below the threshold temperature, the copper conductive sheet is connected. When the temperature has surpassed the threshold temperature, as shown in Figure 3.17(b), the NiTi thermal actuator recovers its memorized position and the conductive sheet is disconnected, activating the tag mode.

Figure 3.17. *Option B. Photograph of the conductive copper sheet before a) and after b) the threshold has been surpassed*

The sensor is read using the time-domain radar (see section 2.3.2) at a 50 cm tag-reader distance. Figure 3.18(a) shows the unprocessed time-domain signal. Figure 3.18(b) shows the signal after background subtraction has been applied. Finally, Figure 3.18(c) shows the signal after both background subtraction and the continuous wavelet transform have been applied. As can be observed, by monitoring the tag mode, the reader can easily detect whether the temperature threshold has been surpassed.

3.2.3.2. *Threshold detector using a thermal switch*

In this case, the sensor is designed using a chipless time-coded UWB tag loaded with a commercial thermal switch. When the temperature exceeds a certain threshold, the impedance of the thermal switch changes. This change produces a modulation on the amplitude of the backscattered pulse in time domain.

A photograph of the tag is shown in Figure 3.19(a). The tag is manufactured on a Rogers RO4003C substrate (see Table 2.3). The tag is composed of a Vivaldi antenna with a slot to microstrip line transition, connected to a delay microstrip line (see section 2.5.2). At the end of the line, a commercial threshold thermal switch model AIRPAX 67F050 is connected. From the manufacturer specifications, a 67F050 thermostat will close (make contact) when temperature surpasses the threshold of 50 ± 5 °C. The antenna is approximately well matched (return loss lower than −10 dB) from 2 to 11 GHz. The thermal switch is characterized to verify its operation in the radar band (3.1–5.6 GHz). To this end, the switch is connected to a short microstrip line and to a PCB SMA connector. The input reflection coefficient of the switch is shown in Figure 3.19(b) after the connector is disembedded. When the switch is cold (temperature below the threshold), it can be modeled as an OC. When it is hot (temperature above the threshold), the two arms of the switch are connected, behaving as a short circuit. At frequencies higher than 5 GHz, the difference between the two states is smaller due to the parasitic effects. Although the switch is not designed to operate at high frequencies, it still works near the radar band.

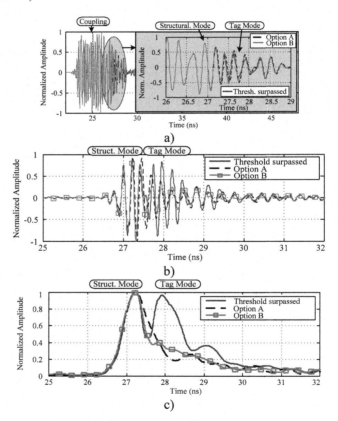

Figure 3.18. *Measured signal of the three tag states: a) unprocessed, b) after background subtraction and c) after background subtraction and CWT. For a color version of the figure, see www.iste.co.uk/ramos/rfid.zip*

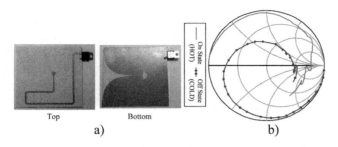

Figure 3.19. *a) Measured reflection coefficient for the two states (On and Off) of the switch between 10 and 6 GHz. b) Photographs of the designed tag (top and bottom). Tag size: 83.43 × 78.4 mm²*

To evaluate the behavior of the switch, the ratio between the energy of the actual differential signal and the maximum energy when the states difference are maximum (e.g. between ideal short and open circuit) is defined. It is the time domain modulation efficiency (η_{mod}) that can be computed in the frequency domain using Parseval's theorem:

$$\eta_{mod} = \int_{-\infty}^{+\infty} |P_{am}(f)\Delta\Gamma_L(f)|^2 \, df \, / \int_{-\infty}^{+\infty} |2P_{am}(f)|^2 \, df \,, \qquad [3.4]$$

where $P_{am}(f)$ is the Fourier transform of the tag mode pulse $p_{am}(t)$. It is found that $\eta_{mod} = 18\%$ for the switch states (from Figure 3.19(b)). To improve this value, the center frequency of the radar should be decreased. Another option would be to increase the sensor operation frequency by integrating it on board.

In order to measure the real temperature as a reference, a PTS (PTS100) [VIS 14] is placed on the thermal dissipator of the switch. To do this, a multimeter that measures the resistance of the PTS is used, similarly as it is performed in section 3.2. Then, the switch is heated with a heatgun. Several acquisitions have been performed at a tag-reader distance of 50 cm. The signal coupled between Tx and Rx antennas at the reader is removed by time gating, that is by selecting the start delay of the acquisition window of the radar (see section 2.4.1). Figure 3.20(a) shows the output of the CWT detector as a function of both the radar time delay and the PTS temperature.

Figure 3.20(b) shows the amplitude at the output of the CWT detector for all the acquisitions as a function of the time delay. The background (measurement of the scene without the tag) has been subtracted to reduce the clutter. For all the acquisitions, it is observed that the structural mode remains constant. The tag mode depends on the reflection at the end of line, which is loaded by the switch. The structural mode amplitude of the Vivaldi antenna, when it is well oriented, is lower than the tag mode. The tag mode changes when the temperature reaches the threshold (about > 50°C). The threshold temperature is within the typical values given by the switch manufacturer (50 ± 5°C). The threshold can be adjusted by changing the model of the switch within the product series in steps of 10°C from 40 to 100°C. Figure 3.20(b) demonstrates that it is possible to determine the switch state by using a simple level comparator at the output of the CWT.

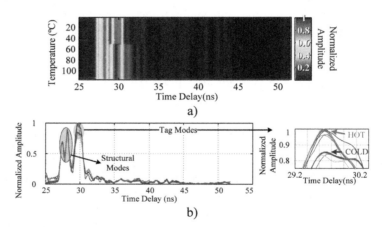

Figure 3.20. *a) Amplitude of the CWT as a function of the time delay and temperature. b) Normalized amplitude of the CWT from the optimum scale as a function of the time delay for hot and cold states. For a color version of the figure, see www.iste.co.uk/ramos/rfid.zip*

3.2.3.3. *Detection techniques for threshold detectors*

When measuring a threshold detector, only two values of the tag mode are required (if the threshold has been surpassed or not).

Therefore, more powerful detection techniques can be applied based on this assumption, since high resolution in amplitude is not required, such as with a linear resistive sensor (see section 3.2.2).

The signal processing techniques used to detect the temperature state are summarized in Figure 3.21. A differential detector is used to detect the change transition in the switch and to mitigate the interferences from the clutter reflections. In order to overcome detection problems at the receiver (associated with a low signal-to-noise ratio when the sensor-reader distance increases), the CWT is used (see section 2.4.2). In addition, a gain compensation function is applied to equalize the attenuation due to propagation loss. Moreover, the delay profile distribution is introduced to determine the range of the tag.

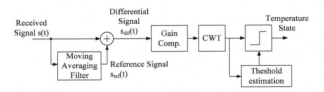

Figure 3.21. *Block diagram of the threshold detection techniques*

First, in order to reduce the effect of non-modulated clutter, the differential signal between two acquisitions is considered:

$$s_{\text{dif}}(t) \approx s(t) - s_{\text{ref}}(t) = \alpha p_{am}(t) \times \delta(t - \tau_P - \tau_L) \times \Delta \Gamma_L(t), \quad [3.5]$$

where $\Delta\Gamma_L(t)$ is the differential reflection coefficient between two acquisitions, $s(t)$ and $s_{\text{ref}}(t)$, and α, as explained in section 2.2, is the attenuation due to the propagation. The delay of the antenna τ_A is removed because it is the same for all acquisitions. If the switch temperature does not change, $\Delta\Gamma_L = 0$. On the contrary, if the threshold is surpassed, $|\Delta\Gamma_L| = |\Gamma_{\text{ON}} - \Gamma_{\text{OFF}}|$. For both cases, the structural mode is suppressed, since it does not change between two different acquisitions. In practice, the reference signal can be obtained as the output of a moving averaging filter in order to reduce noise.

Signals at longer distances (long delays) have smaller amplitudes than those from closer distances (short delays). Therefore, it is needed to equalize amplitudes applying a time gain function, to compensate the attenuation due to the propagation loss (α in [3.5]). α assumes that line-of-sight (LOS) propagation is inversely proportional to the square of distance or time delay. Gain compensation proportional to the square of time delay has been applied in order to compensate α.

After applying the CWT, the next step is to decide a threshold for the temperature state estimation. To this end, the CWT coefficients for the maximum scale (a_m) are recorded in a matrix R_{mn}, where the row index m represents the number of radar acquisition and column index n represents the time delay of the acquired signal. Then, the average delay profile (ADP) is computed as the average of the amplitudes for each row (the threshold is obtained by analyzing the column index n that maximizes the delay profile):

$$P(n) = \frac{1}{N}\sum_{m=1}^{N}|R_{mn}|^2 \ . \tag{3.6}$$

An example of the detection techniques, applied to the temperature threshold detector based on a commercial thermal switch (see section 3.2.3.2) is shown next. Figure 3.22(a) shows the maximum of the CWT for different acquisitions at a 5 m tag-reader distance. Figure 3.22(b) shows the measured signals after the gain compensation has been applied. Finally, undesired clutter is removed using $s_{ref}(t)$, avoiding the use of the background subtraction. $s_{ref}(t)$ is obtained by averaging the first acquisitions. This is possible because the switch is assumed to be cold when the monitorization starts, before the temperature could exceed the threshold. Then, the differential signal is obtained by subtracting each acquisition from the reference signal. Figure 3.22(c) shows the differential signal obtained from the gain-compensated signal from Figure 3.22(b). The temperature state is obtained by comparing the output of CWT with a threshold value.

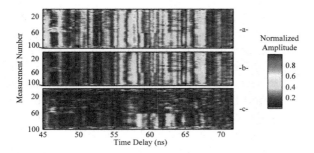

Figure 3.22. *Normalized amplitude of the CWT for the optimum scale before a) and after b) applying gain compensation. c) Normalized amplitude of the CWT of the Differential signal. The distance between reader and tag is 5 m. For a color version of the figure, see www.iste.co.uk/ramos/rfid.zip*

Figure 3.23(a) shows the ADP distribution (see equation [3.6]) of the CWT, obtained from the averaging of all acquisitions for each time delay. The peak corresponds to the maximum variation of the differential signal due to the change in the antenna mode of the sensor. From the time delay of the peak, it is easy to localize the tag distance. A cut of the output of CWT for this time delay is shown in Figure 3.23(b). Figure 3.23(c) shows the complementary distribution function (CDF) of the amplitudes of Figure 3.23(b). According to this function, the threshold is fixed to the midpoint between the minimum and the maximum points. The measurements whose amplitude is higher than this threshold value correspond to hot switch states. In addition, the fraction of time when the detector is cold (T_{COLD}/T) can be obtained from this graphic.

3.2.4. *Self-calibration and reliability*

There are a number of issues that must be addressed in order to make chipless sensors competitive in front of other alternative chip-based solutions. Some of these aspects are the read range, reliability, accuracy or sensor calibration process.

As discussed in section 3.2.2, the backscattered signals in amplitude-based chipless sensor tags only contain information of the physical parameter that is sensed. No other state is available to perform a calibration. This means that the user needs a calibration curve for all possible sensor-reader distances and angles. Two tags could be used for simultaneous measurement, one sensor tag and a calibration tag. This results in a large structure and the problem partly remains, since the tag-reader angle is not always identical for the two tags. Next, a second backscattering signal is integrated at the sensor tag to perform a self-calibration. In addition, this signal also increases the number of words that can be coded to identify the tag.

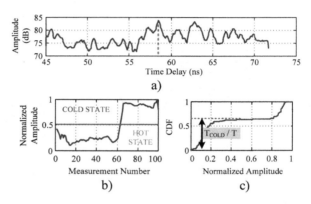

Figure 3.23. *a) Average delay profile distribution (ADP) of the CWT results. b) Normalized amplitude for the delay of the peak of ADP and threshold value used in the state decision. c) CDF plot*

The sensor tag is composed of a UWB Vivaldi antenna (see section 2.5.2). In this case, however, the antenna is connected to two transmission lines with different lengths (L_1 and L_2, where $L_1 < L_2$). A circuit scheme of the tag and its corresponding photograph are shown in Figure 3.24.

One line is terminated with an OC (*Line 1*) and the other is loaded with a resistive temperature sensor (*Line 2*) that modulates its reflection coefficient Γ as a function of the temperature ($\Gamma = f(T)$). The reader sends a pulse $p(t)$, and receives the signal backscattered by the tag. The received time-domain signal is mainly composed of three reflected pulses: a structural mode (which depends on the tag size, shape and material) and two tag/antenna modes (which depend on the load connected at each of the two transmission lines). The *Tag1* mode is used to normalize the *Tag2* mode, instead of normalizing *Tag2* with the structural mode (see section 3.2.2).

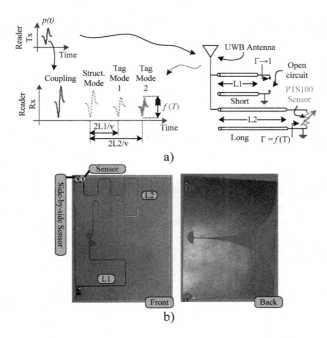

Figure 3.24. *Scheme of the signals transmitted and received at the reader and tag circuit scheme: a) tag photograph of front face b, left) and back face (b, right). The tag is manufactured on Rogers 4003 substrate. Tag size: 11.5 cm × 8.7 cm. For a color version of the figure, see www.iste.co.uk/ramos/ rfid.zip*

As explained in section 2.6.3, the angular behavior of the structural mode and tag mode is different. The structural mode depends on the RCS associated with the shape and materials of the tag. The tag mode is associated with the reradiated fields on the antenna (it follows the radiation pattern of the antenna). In consequence, the structural-to-tag mode ratio depends on the illumination angle, whereas the *Tag1*-to-*Tag2* modes ratio is independent of the angle because the two amplitudes depend on the same radiation pattern.

Figure 3.25 shows the measured tag response as a function of several resistive loads connected at the ends of lines L_1 and L_2. The processing techniques from section 2.4 have been applied. The first peak corresponds to the structural mode. The second and third peaks correspond to the tag modes for the (short-delay) line L_1 (*Tag1*) and for the (long-delay) line L_2 (*Tag2*), respectively. The fourth and next peaks are multiple reflections of the tag modes. When the lines are terminated with an OC, their associated tag modes have the largest amplitude. On the other hand, when the loads are matched to the characteristic impedance of the lines (80 Ω) their amplitudes are very small.

Figure 3.26 shows the measured tag response when the load connected to the line L_2 is changed from 82 to 180 Ω. It can be observed that while the structural and *Tag1* modes remain invariant, the *Tag2* mode changes its amplitude. Figure 3.27 shows the measured ratio between the structural mode and the *Tag2* mode, and between the *Tag1* and *Tag2* modes as a function of the load connected at the end of line L_2. It can be observed that both ratios depend on the load. Therefore, the *Tag1*-to-*Tag2* ratio can be used to detect the load at L_2.

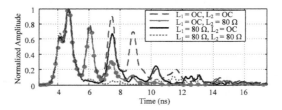

Figure 3.25. *Structural modes (first peak), tag modes of the short-delay line L_1 (second peak, Tag1) and tag modes of the long-delay line L_2 (third peak, Tag2) depending on the loads connected at their ends. For a color version of the figure, see www.iste.co.uk/ramos/rfid.zip*

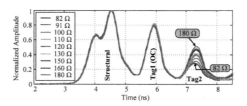

Figure 3.26. *Tag response as a function of the load at the end of line L_2. For a color version of the figure, see www.iste.co.uk/ramos/rfid.zip*

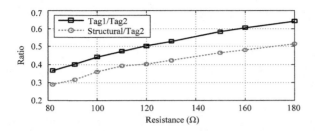

Figure 3.27. *Ratio between Tag1 and Tag2 modes and between the structural mode and Tag2 mode as a function of the load, measured at a 40 cm tag-to-reader distance. For a color version of the figure, see www.iste.co.uk/ramos/rfid.zip*

Figure 3.28(a) shows the structural mode and Figure 3.28(b) shows *Tag1* and *Tag2* modes as a function of the tag-reader angle in H-plane, being the *Tag2* mode loaded with 180 Ω. The structural mode can be seen as the RCS of the tag, and the tag modes as the tag radiation pattern over the entire frequency band. Taking into consideration the angles where the antenna radiates (−30° to 30°), Figure 3.29 shows the structural-to-*Tag2* and the *Tag1*-to-*Tag2* ratios. It can be deduced that using the ratio between tag modes is more constant than using the structural mode. Both ratios have been empirically demonstrated to be fairly constant with distance in section 2.6.1.

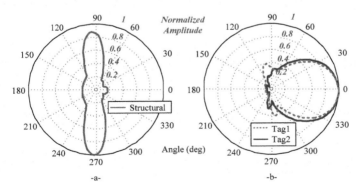

Figure 3.28. *a) Structural mode and b) tag modes as a function of the tag-reader angle in H-plane*

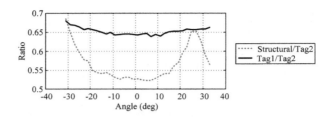

Figure 3.29. *Ratio between structural and Tag2 modes and between Tag1 and Tag2 modes for the −30° to 30° range of Figure 3.28(b)*

Next, a Vishay PTS 100 Ω temperature sensor is connected at the end of line L_2. A second identical sensor is placed side-by-side in order to measure the resistance with a multimeter for characterization purposes (see Figure 3.6). The reflection coefficient at the end of line L_2 changes with the temperature and so does the amplitude of *Tag2* mode. In this section, it is shown the advantage of obtaining the temperature from the ratio between *Tag1* and *Tag2* modes in front of the ratio between the structural and *Tag2* modes. Figure 3.30 shows the measured response when the sensor temperature is changed from 36 to 130°C. This measurement has been done using the step-frequency approach (see section 2.3.1).

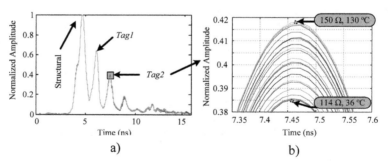

Figure 3.30. *a) Measured tag response as a function of the temperature with the step-frequency technique at 40 cm and b) zoomed response of the Tag2 mode. For a color version of the figure, see www.iste.co.uk/ramos/rfid.zip*

Figure 3.31 shows the estimated temperature from the tag mode as a function of the real temperature for the structural-to-*Tag2* ratio and the *Tag1*-to-*Tag2* ratio. The temperature is obtained from the resistance measurement with the side-by-side sensor and the multimeter. The processing techniques described in section 3.2.2 have been performed. It can be observed that using the *Tag1*-to-*Tag2* ratio clearly reduces the measurement error.

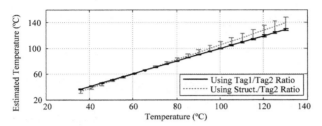

Figure 3.31. *Estimated temperature as a function of the temperature using the ratio between Tag1 and Tag2 modes (solid black line) and using the ratio between the structural and Tag2 modes (dashed red line) measured with the step-frequency technique at 40 cm*

Finally, Figure 3.32 shows the same measurement done with the time-domain radar (section 2.3.2). An identical behavior is observed, although the measurement error is larger due to the higher noise of its receiver in front of using a VNA.

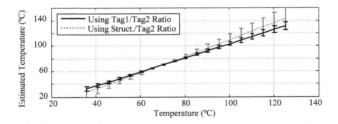

Figure 3.32. *Estimated temperature as a function of the temperature using the ratio between Tag1 and Tag2 modes (solid black line) and using the ratio between the structural and Tag2 modes (dashed red line) measured with the impulse technique at 40 cm*

3.3. Delay-based time-coded chipless sensors

Similarly, as seen in section 3.2, it is possible to sense a physical parameter that changes, in this case, the delay of the tag mode. If a material with a specific permittivity is added on top of the transmission line of a time-coded chipless UWB tag, the permittivity of the medium changes, and therefore the tag mode delay also changes. The main idea is to create a wireless permittivity sensor based on this principle. In this context, a wireless permittivity sensor using chipless frequency-coded UWB RFID is presented in [GIR 12], detecting dielectric permittivities between 1 and 4.3 at 70 cm of distance.

There are several application fields where it is interesting to measure permittivity wirelessly. Capacitive sensors are used for monitoring several magnitudes such as humidity and barometric pressure (this last magnitude depends on the air permittivity). These sensors are based on the variation of the capacitance due to change in the permittivity, as a function of the physical or chemical magnitude. Moreover, variations in the complex dielectric constant can be useful to detect soil composition or variations in civil materials composition, such as the mixture of construction mortars. Aging and long-term evolution of the structure is a major topic in concrete-based buildings [MOR 93]. Construction aggregate (sand, gravel and rocks) and cement have different permittivities, depending on the percentage of each element. This will be studied in detail in section 3.3.2.2.

3.3.1. *Principle of operation*

Figure 3.33 shows a scheme of the chipless UWB time-coded permittivity sensor and the signals between the reader and tag. The permittivity of the medium (transmission line + material added on top) is denoted as ε_{eff}. The structural-to-tag mode delay depends on the permittivity of the material. Moreover, the tag mode amplitude depends on the loss of the medium, which, in turn, depends on the loss tangent of the material $tan\delta_2$ (the permittivity of the substrate is known).

The permittivity of the medium ε_{eff} can be calculated using the conformal mapping method proposed in [SVA 92]. In this case, a two-layer structure is considered, where ε_{r1} is the tag substrate permittivity, ε_{r2} is the material permittivity, w is the transmission line width, h is the substrate thickness and h_2–h is the material thickness. ε_{efr} can be expressed as:

$$\varepsilon_{\text{efr}} = \varepsilon_{r1}q_1 + \varepsilon_{r2}\frac{(1-q_1)^2}{\varepsilon_{r2}(1-q_1-q_2)+q_2}.$$ [3.7]

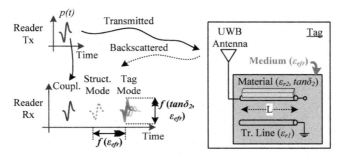

Figure 3.33. *Scheme of the wireless chipless permittivity sensor and signals between the tag and the reader*

If $w/h > 1$, then the filling factors q_1 and q_2 are:

$$q_1 = 1 - \frac{1}{2}\frac{\ln\left(\frac{\pi}{h}w_{ef}-1\right)}{w_{ef}/h},$$ [3.8]

$$q_2 = 1 - q_1 - \frac{1}{2}\frac{h-v_e}{w_{ef}}\ln\left[\pi\frac{w_{ef}}{h}\frac{\cos\left(\frac{v_e\pi}{2h}\right)}{\pi\left(\frac{h_2}{h}-\frac{1}{2}\right)+\left(\frac{v_e\pi}{2h}\right)}+\sin\left(\frac{v_e\pi}{2h}\right)\right],$$ [3.9]

where w_{ef} and v_e are:

$$w_{ef} = w + \frac{2h}{\pi}\ln\left[17.08\left(\frac{w}{2h}+0.92\right)\right],$$ [3.10]

$$v_e = 2\frac{h}{\pi}arctg\left[\frac{\pi}{\pi w_{ef}/2h-2}\left(\frac{h_2}{h}-1\right)\right].$$ [3.11]

If $w/h < 1$, the filling factors q_1 and q_2 are:

$$q_1 = \frac{1}{2}\left[1 + \frac{\pi}{4} - \frac{1}{2}\arccos\left(\frac{w}{8h}\right)\right]\sqrt{\frac{8h}{w}}, \qquad [3.12]$$

$$q_2 = \frac{1}{2} - \frac{0.9 + \frac{\pi}{4}\ln(b)\arccos\left\{\left[1 - \frac{h}{h_2}\left(1 - \frac{w}{8h}\right)\right]\sqrt{b}\right\}}{\pi\ln(8h/w)}, \qquad [3.13]$$

where b is defined as:

$$b = \frac{(h_2/h) + 1}{(h_2/h) + (w/4h) - 1}. \qquad [3.14]$$

Figure 3.34 shows the calculated medium permittivity for the $w/h < 1$ and the $w/h > 1$ cases. An $\varepsilon_{r1} = 3.55$ and $h = 0.813$ mm have been chosen, which correspond to a Rogers RO4003C substrate (see Table 2.3). As it can be observed, for both cases the permittivity remains practically constant for material heights greater than 4 mm. Therefore, in practice the medium permittivity will depend only on ε_{r2}. It can also be observed that ε_{eff} is more sensitive for the $w/h < 1$ case. Therefore, the tag should be designed with a ratio $w/h < 1$.

For the case of interest ($w/h < 1$), the characteristic impedance Z_0 of the multilayer transmission line can be expressed as:

$$Z_0 = \frac{60}{\sqrt{\varepsilon_{efr}}}\ln\left(\frac{8h}{w}\right), \qquad [3.15]$$

and it is plotted in Figure 3.35, for a $w = 0.7$ mm. It can be observed how the change in the material permittivity mismatches the impedance between the antenna (designed to be an impedance of 83 Ω) and the transmission line attached to the material. Therefore, the tag mode amplitude will depend both on the loss tangent and the permittivity of the material and no change is observed for material heights greater than 4 mm.

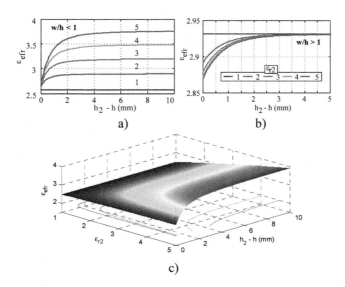

Figure 3.34. *Calculated ε_{efr} for the a) w/h < 1 case and b) w/h > 1 case. c) Detail of the case where w/h < 1 with a continuous sweep. The material height h is swept from 1 to 10 and its permittivity ε_{r2} from 1 to 5. For a color version of the figure, see www.iste.co.uk/ramos/rfid.zip*

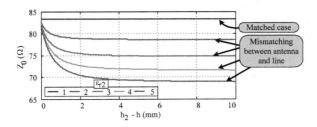

Figure 3.35. *Impedance of the medium (transmission line) as a function of the permittivity and height of the material added on top of it. For a color version of the figure, see www.iste.co.uk/ramos/rfid.zip*

3.3.2. *Permittivity sensor based on chipless time-coded UWB tags*

Two tag approaches are designed and simulated for chipless UWB time-coded permittivity sensors; both on Rogers RO4003C substrate (see Table 2.3). Then, the tag with the best performance is used in a real application to detect concrete composition.

3.3.2.1. *Sensor design*

The first permittivity sensor tag consists of the small monopole discussed in section 2.6.2, but the delay line length L has been increased to increase time resolution. By increasing L, there will be a greater change in $\Delta\tau_L$ due to the attached material. Five thick slabs of materials are considered, with a thickness of 1 cm, which is much larger than the 4 mm required (see section 3.3.1). The materials are characterized and their respective permittivities are: PTFE (2.2), PC (3.2), PET (3.6), PUR (4) and CARP (5.7). Figure 3.36(a) shows the layout of the designed tag. The tag size is 34.5 mm × 69 mm. Figure 3.36(b) shows the simulated $|S_{11}|$ parameter. Finally, Figure 3.36(c) shows the designed tag with one material attached on top of the transmission line in the simulation environment.

The response of the tag when in contact with the materials is simulated using Ansys HFSS, with a plane wave orientated toward the tag bottom face, and a vacuum box to set a distance of 40 cm. The plane wave is oriented at the back face for simplicity. Otherwise, the contribution of the material would appear before the contribution of the structural mode of the tag.

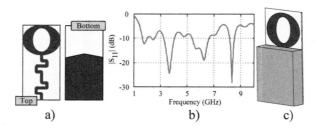

Figure 3.36. *a) Layout of the tag. b) Simulated $|S_{11}|$.*
c) Scheme of the simulated tag with attached material

Figure 3.37 shows the time-domain response of the tag before and after applying the CWT technique discussed in section 2.4.2. As it can be observed, the structural mode remains invariant for all the measurements, while the tag mode delay increases when the permittivity of the material increases.

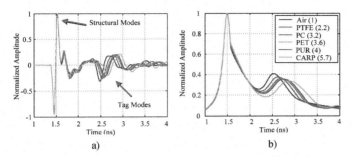

a)   b)

Figure 3.37. *Simulated time-domain response of the tag with several materials attached on top of the transmission line. a) RAW signal. b) Continuous wavelet transform has been applied to the RAW signal. For a color version of the figure, see www.iste.co.uk/ramos/rfid.zip*

Figure 3.38(a) shows the structural-to-tag mode delay as a function of the permittivity of the material attached to the transmission line. A linear behavior can be observed. Figure 3.38(b) shows the delay increase with respect to air ($\varepsilon_{r2} = 1$) case. Again, a linear behavior is observed, with a delay increase from 100 to 350 ps.

Finally, Figure 3.39 shows the structural-to-tag mode ratio as a function of the permittivity of the material. The material reduces the tag mode amplitude. It will be studied in detail next.

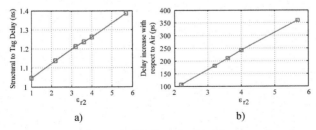

a)   b)

Figure 3.38. *a) Structural-to-tag mode delay as a function of the permittivity of the material attached to the tag transmission line, b) Increase in delay with respect to the air ($\varepsilon_{r2} = 1$) case*

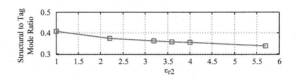

Figure 3.39. *Structural-to-tag mode ratio as a function of the permittivity of the material attached to the tag transmission line*

The second permittivity sensor tag consists of a Vivaldi antenna (see section 2.5.2) connected to a meandered transmission line with width $w = 0.7$ mm (about $Z_0 = 83$ Ω, as used in section 3.3.1). The tag layout and its dimensions are shown in Figure 3.40(a). It has a size similar to a standard credit card. As explained in section 2.5.2, longer (with respect to small monopoles) delay line lengths L can be integrated increasing the sensitivity of the system. Figure 3.40(b) shows the simulated (using Agilent Momentum) $|S_{11}|$ parameter of the tag.

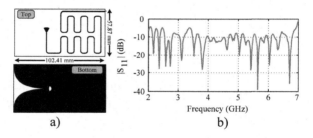

a)   b)

Figure 3.40. *a) Designed tag layout and dimensions. b) Simulated $|S_{11}|$ parameter of the tag*

The tag has been simulated using Ansys HFSS. The material height is $h_2 - h = 10$ mm. A plane wave is sent to the tag and the backscattered response is simulated up to 10 ns. Figure 3.41(a) shows the simulated RAW time-domain signal for a material permittivity $\varepsilon_{r2} = 1.3$ and a loss tangent $tan\delta_2$ from 0 to 0.29. The structural and tag modes can be seen. In the inset, the tag mode is shown in detail. Figure 3.41(b) shows the same signal after applying the CWT. Finally, Figure 3.41(c) shows the same processed signal as in Figure 3.41(b), but with a block of permittivity $\varepsilon_{r2} = 5$. The tag mode delay is changed from about 6.5 to 7 ns, while the tag mode amplitudes are also reduced with respect to the $\varepsilon_{r2} = 1.3$ case.

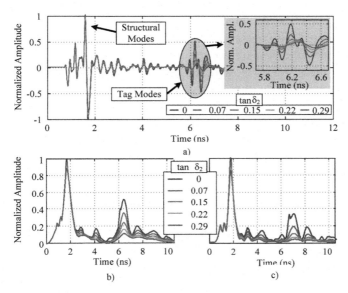

Figure 3.41. *Simulated, a) RAW and b) processed time-domain signal for the tag with an $\varepsilon_{r2} = 1.3$ block and a varying $\tan\delta_2$ between 0 and 0.29, and c) with an $\varepsilon_{r2} = 5$ block. For a color version of the figure, see www.iste.co.uk/ramos/rfid.zip*

In order to interpret this information, Figure 3.42 shows the tag mode amplitude as a function of both the material permittivity and the loss tangent. It can be observed that, for $\varepsilon_{r2} > 3$, the tag mode amplitude change mainly depends on the loss tangent of the material.

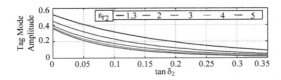

Figure 3.42. *Simulated tag mode amplitude as a function of the material permittivity and loss tangent. For a color version of the figure, see www.iste.co.uk/ramos/rfid.zip*

Next, the tag is simulated attaching a 10 mm-thick material with no loss and varying the permittivity, similarly as with the small monopole case. The tag mode delay changes according to the permittivity, while the tag mode amplitude depends on the mismatching between the UWB antenna and the medium with the material. Figure 3.43 shows the simulated, processed time-domain response. The signal is normalized with respect to the absolute maximum, which is the structural mode for $\varepsilon_{r2} = 1$. In this case, the structural mode amplitude varies depending on the permittivity. This is because the Vivaldi antenna has a smaller RCS than a UWB monopole. The absorption due to the material affects the structural mode amplitude. Figure 3.44 shows the amplitude of the tag mode as a function of ε_{r2}. The same behavior as shown in Figure 3.42 is observed. For permittivities $\varepsilon_{r2} > 3$, the tag mode amplitude does not vary. Finally, the tag mode delay is shown in Figure 3.45 as a function of ε_{r2}. As with the small monopole case (see Figure 3.38), a linear behavior is observed.

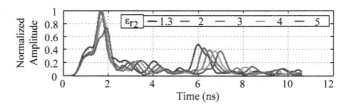

Figure 3.43. *Simulated, processed time-domain signal for the tag with a non-lossy material with permittivities from 1 to 5. For a color version of the figure, see www.iste.co.uk/ramos/rfid.zip*

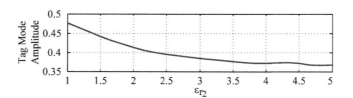

Figure 3.44. *Tag mode amplitude as a function of ε_{r2}, with $\tan\delta_2 = 0$ (lossless material)*

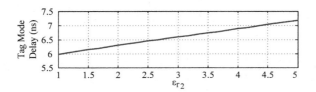

Figure 3.45. *Tag mode delay as a function of ε_{r2}*

3.3.2.2. *Concrete composition detection*

The structural health of concrete-based civil structures is a major concern in today's society. Detecting the quality of the materials used in civil engineering is necessary to ensure safe structures, such as buildings or bridges [LYN 07]. Unexpected events such as earthquakes, hurricanes, or simply a material deterioration because of a wrong mixture of compounds, could cause a structure to collapse. During the construction period, samples of the construction compounds are usually inspected. However, long-term non-destructive testing (NDT) of these structures is also desired. In this context, several works have been presented recently, with the aim being to wirelessly sense these structures. Battery-powered active sensors, using microcontrollers and accelerometers, are reviewed in [LYN 07]. One of the main materials used in this type of structures is concrete, specifically a mixture of concrete and dry sand. The permittivity of these two materials has been studied at the microwave frequency range. A passive permittivity sensor that uses UHF RFID at 870 MHz is presented in [MAR 03]. A UHF RFID sensor is presented to detect the permittivity of lightweight concrete in [SUW 14]. The dielectric constant of concrete is obtained in [HUI 08] by measuring the deflection and loss of electromagnetic waves in concrete blocks using a CW radar. In [RHI 98], the permittivity of concrete is measured between 0 and 20 GHz using a probe and a network analyzer. In [MAT 98], the permittivity of dry sand is measured up to 6 GHz using microwave resonators. A wireless water content sensor based on a 22.5 MHz inductive-capacitive circuit is presented in [ONG 08].

Next, the Vivaldi-based permittivity sensor tag will be used as a batteryless concrete composition (concrete and sand) sensor. In addition, a method to calibrate the tag is presented to obtain the concrete composition.

Figure 3.46(a) shows a scheme of the system. The delay line is embedded in concrete, as shown in both the schemes of the tag (Figure 3.46(b)), layers (Figure 3.46(c)) and photographs (Figure 3.46(d)). The signal at the reader's receiving (Rx) antenna in time domain can be approximated as the sum of several peaks (see section 2.2):

$$s(t) = s_{\text{Coupl.}}(t) + s_{\text{Clutter}}(t) + s_{\text{Str.}}(t) + s_{\text{Plate}}(t) + s_{\text{Concrete}}(t) + s_{\text{Tag}}(t), \quad [3.16]$$

where $s_{\text{Coupl.}}$ is the coupling from the reader Tx to Rx antenna and s_{Clutter} is the clutter from the scene. Both can be removed by windowing and background subtraction (see section 2.4.1). $s_{\text{Str.}}$ corresponds to the tag structural mode, which depends on the tag shape, size and material. A Vivaldi antenna has a small RCS in its optimum radiation pattern H-plane angle. Therefore, its structural mode is small and difficult to detect. A metal plate is soldered in the Vivaldi ground plane in perpendicular to the Vivaldi ground plane, providing a strong reference peak. This plate is separated from the tapered transition so it does not affect the antenna. s_{Plate} corresponds to the signal backscattered at the metal plate, s_{Concrete} corresponds to the reflection at the concrete slab or wall. Finally, s_{Tag} corresponds to the tag mode, which depends on the load of the antenna (here an open-ended delay line). s_{Tag} is the part of $p(t)$ that propagates inside the tag and is reradiated to the reader with the information of the material where the line is embedded.

The Vivaldi antenna has been chosen because it permits to integrate a long meander delay without having undesired effects, as studied in section 2.5.2. Also, its gain has shown to be larger than the other studied alternatives, and its boresight is perpendicular to the concrete wall or slab. As shown in Figure 3.46(a), the signals corresponding to $s_{\text{Str.}}$ and s_{Concrete} are unstable in time, because of changes in angle or distance from one measurement to another, and because of dispersion, respectively. Therefore, s_{Plate} is used as reference, which also denotes the limit where the tag should be embedded in concrete. The delay between s_{Plate} and s_{Tag} is addressed as τ. τ depends on the delay line length L and the propagation speed of the medium v, which, in turn, depends on the effective permittivity of the medium ε_{eff}. ε_{eff}, in turn, depends on the permittivities of both the substrate (ε_{r1}) and the concrete slab (ε_{r2}). Finally, the term $2d/c$ ($c = 3 \times 10^8$ m/s) accounts for the distance d between the tag and reader, from which τ is independent.

3.3.2.3. Results

In order to obtain the concrete layer permittivity, a relationship between the effective permittivity of the multilayer microstrip and the concrete layer permittivity is needed. The S parameters of a multilayer microstrip transmission line of width $w = 0.7$ mm on RO4003C (see Table 2.3) have been simulated with Agilent Momentum. The structure follows the scheme of Figure 3.46(c), the concrete layer height is $h_2 - h = 10$ mm and ε_{r2} is swept from 1 to 5. A line of length $L = 2.5$ mm is chosen to minimize the time required for the simulation. The effective permittivity ε_{eff} of the multilayer structure is obtained from the expression of S parameters of a transmission line. Then:

$$\gamma_l = \text{acosh}\left(\frac{Z_{11}}{Z_{21}}\right),$$ [3.17]

$$\beta = \text{Im}\left(\frac{\gamma_l}{L}\right),$$ [3.18]

$$\varepsilon_{\text{eff}} = \left(\frac{\beta c}{2\pi f}\right)^2,$$ [3.19]

where Z_{11} and Z_{21} are the simulated Z parameters of the multilayer transmission line, and f is the frequency. Since the reader is based on the time-domain radar (see section 2.3.2), which is centered at 4.3 GHz, the permittivity ε_{eff} at this frequency is chosen for each simulated ε_{r2} case. Figure 3.47 shows the resulting curve. A nearly linear behavior is obtained, as the one observed in Figure 3.34(c) for thicknesses greater than 4 mm.

The tag with the concrete block (see Figure 3.46(b)) has been simulated using Ansys HFSS. A 10 mm-thick concrete block is added on top of the tag delay line. The permittivity of this block (ε_{r2}) is again swept from 1 to 5, with 20 samples within the range. Figure 3.48 shows the simulated time-domain signal before and after applying the CWT. It is important to note that not all the samples of ε_{r2} are shown for representation convenience. On the contrary as the case simulated in Figure 3.43, in this case the metal plate fixes the time reference, while the tag mode varies depending on ε_{r2}.

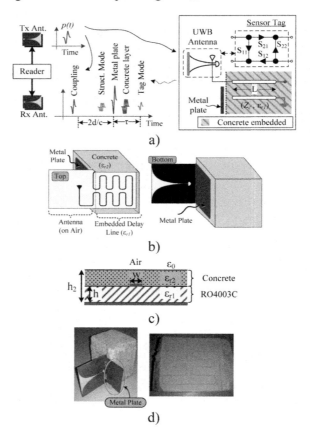

Figure 3.46. *a) Scheme of the tag-reader system. b) Scheme of the tag embedded in concrete. c) Scheme of the layers in the tag. d) Photograph of the tag embedded in concrete (left) and cut of the concrete layer (right). For a color version of the figure, see www.iste.co.uk/ramos/rfid.zip*

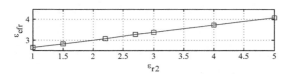

Figure 3.47. *Simulated effective permittivity of the multilayer structure*

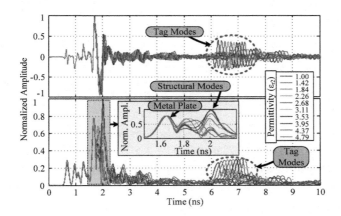

Figure 3.48. *Simulated time-domain signal of the tag for ε_{r2} from 1 to 5: RAW signal (top); signal after applying the CWT (bottom). For a color version of the figure, see www.iste.co.uk/ramos/rfid.zip*

Three concrete blocks (A, B and C) are fabricated with an embedded tag. The blocks are composed of a mixture of portland concrete and sand. The compositions are depicted in Table 3.2. The blocks have been dried in still air for 4 weeks before measuring them. Three sets of 100 measurements have been performed at a 40 cm tag-reader distance. Figure 3.49 shows the measured time-domain response of the blocks after applying the CWT. The background subtraction technique (see section 2.4.1) is also applied. In the case of concrete walls, the background can be obtained from a tag-free portion of the wall at the same distance. The mean delays $\overline{\tau}$ between the metal plate and the tag modes are shown in Table 3.2. The tag mode amplitude is reduced when the tag is embedded in concrete but the peaks and the time difference can still be perfectly detected. The measured delay τ can be modeled as:

$$\tau = \tau_0 + 2L_{\text{eff}}\sqrt{\varepsilon_{\text{eff}}}\,/\,c, \qquad\qquad [3.20]$$

where τ_0 accounts for a constant delay offset between the plate and the delay line, and L_{eff} is the equivalent electrical delay line length. τ_0 and L_{eff} are unknown, and have to be experimentally determined as calibration parameters. The procedure works as follows:

1) the tag is measured in free space ($\varepsilon_{r2} = 1$) and with an attached 10 mm PTFE slab (known stable material, $\varepsilon_{r2} = 2.2$);

2) the equivalent ε_{eff} for free space and PTFE are obtained from the simulations in Figure 3.47;

3) using the delays from 1), equivalent permittivities from 2), and equation [3.21], $L_{\text{eff}} = 0.27$ m and $\tau_0 = 1.571$ ns. These parameters are stored as calibration for these tags and setup, and work at any distance;

4) using the calibration parameters from 3) and $\bar{\tau}$ from Table 3.2, ε_{eff} can be obtained, and ε_{r2} is derived by inverting Figure 3.47.

The permittivity of dry sand is between 2.5 and 2.75 [MAT 98]. The permittivity of concrete is around 5 [SUW 14, HUI 08]. Therefore, the estimated ε_{r2} falls within the expected range for the mixtures. To validate these measurements, the permittivities of the samples are also measured using a microstrip circular ring resonator (CRR) [BER 91] and an Agilent E8364C Vector Network Analyzer. These results are also shown in Figure 3.50, as well as in Table 3.2. The internal radius of the resonator is $r = 11.46$ mm. The permittivity is then obtained as:

$$\varepsilon_{\text{eff,CRR}} = \left(\frac{c}{Lf_r}\right)^2 = \left(\frac{c}{2\pi r f_r}\right)^2, \qquad\qquad [3.21]$$

where f_r is the resonant frequency (peak). A good agreement between the wireless measurements and the CRR reference is observed. It is important to note that, because of the radar time resolution ($\pm$ 30 ps), there is an error in ε_{r2} about $\pm$ 0.1. As shown in equation [3.20], the permittivity depends on a square root term. Therefore, there is more sensitivity for smaller permittivities, that is, when the part of sand increases.

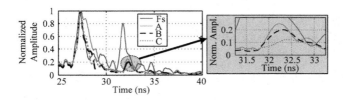

Figure 3.49. *Measured time-domain response of the tag embedded in three concrete blocks and on air after applying the CWT. For a color version of the figure, see www.iste.co.uk/ramos/rfid.zip*

Block number	% C	% S	Delay $\bar{\tau}$ (ns) $\pm \sigma$	Estimated ε_{r2}	Ref. ε_{r2} CRR
A	25	75	5.000	3.56	3.51
B	50	50	5.070	3.98	4.07
C	100	0	5.125	4.31	4.24

% C, percentage of concrete; % S, percentage of sand, σ = 30 ps.

Table 3.2. *Concrete blocks composition*

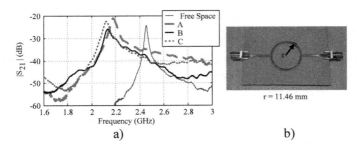

Figure 3.50. *a) Simulated $|S_{21}|$ parameter of a microstrip circular ring resonator with the samples on top. b) Photograph of the circular ring resonator. For a color version of the figure, see www.iste.co.uk/ramos/rfid.zip*

3.4. Conclusions

This chapter has presented the realization of several chipless UWB time-coded sensors. Two main approaches have been studied: amplitude-based sensors and delay-based sensors. The following conclusions have been drawn:

– it is possible to realize chipless temperature sensors by using a simple platinum resistor at the end of a chipless time-coded UWB tag to detect temperature in the 30–130°C range;

– the sensor provides the temperature from the tag mode amplitude of the backscattered signals and the tag is identified by the delay of the line connected to the antenna;

– reliable measurements up to a distance of 1 m have been obtained, which is a smaller distance than the 2 m for just identification (see section 2.6.1);

– a characterization of the system as a function of the angle between the reader and the tag sensor has also been carried out. A good performance for all angles, except for the radiation nulls of the UWB antenna, is demonstrated;

– calibration is a major concern to obtain a reliable measurement. With a chipless tag, which generates two simultaneous states, it can be achieved. It permits to make the sensor independent from the illumination angle within the beamwidth of the tag antenna;

– if it is only needed to measure whether a threshold has been surpassed, more advanced processing techniques can be applied, increasing the reading distance from 1–2 m up to 5 m;

– concrete composition detection can also be remotely measured using very inexpensive chipless tags. This is desirable in civil engineering for classification purposes and long-term quality evaluation, where the number of sensors is expected to be very large.

Semi-passive Time-coded UWB RFID: Analog and Digital Approaches

4.1. Introduction

There is a growing interest in compact, autonomous devices that can monitor the environment. Chipless sensors, as discussed in Chapter 3, provide a low-cost alternative for certain application fields, where high accuracies are not needed. There are, however, more advanced applications in the Internet of things (IoT) where high accuracy and reliability are needed, as well as a large number of sensor IDs [WEL 09, KOR 10]. UWB-based RFID can be an enabling technology for the IoT, specifically in the link between each of the vast number of scattered sensors and readers. In these applications, more advanced approaches than chipless RFID are needed. For instance, in [PRE 07], a semi-passive transponder at 2.4 GHz using a commercial microcontroller is presented. In [YIN 10], a standard Gen2 UHF tag has been integrated with a temperature sensor. In [VYK 09], a custom approach using UHF and microcontrollers is presented using paper substrates.

This chapter presents a semi-passive time-coded UWB platform for remote identification and sensing. The following two approaches are presented:

– a digital, microcontroller-based approach, where the tag backscatters binary information;

– an analog approach, where the tag changes its state between continuous values in RF.

A general scheme of the semi-passive time-coded UWB RFID systems is shown in Figure 4.1. It comprises a reader and the tags. The reader interrogates the tag using a 2.45 GHz signal and an independent narrowband antenna, which activates the tag circuitry. The frequency band of 2.4–2.5 GHz belongs to the ISM bands of free domestic use [LOY 14]. Therefore, the 2.4-2.5 GHz ISM frequency is chosen for this link. For longer distances, the UHF RFID frequency bands can be used because higher transmitted power is allowed. The signal of 2.45 GHz is modulated in order to prevent false wake-ups due to other communication systems near the tag. Once the tag is woken by the reader, the reader sends a UWB pulse to the tag, which hits the tag and is then backscattered toward the reader. This backscattered answer is modulated by the tag, according to the information that it wants to answer. The main parts which comprise the system are as follows:

– a wake-up interrogation system to activate the tag. The tag is normally in a low-power (sleep) mode;

– the core circuitry of the tag. Depending on the approach (analog or digital) it can be;

- a low-power digital microcontroller,

- analog circuitry, mainly consisting of operational amplifiers, passive elements and DC switches,

– a UWB time-coded backscatterer, which is modulated by the core circuitry.

In all cases, the reader is implemented with the time-domain UWB radar (see section 2.3.2).

The chapter is organized as follows:

– section 4.2 presents the wake-up system, based on a Schottky diode rectifier, used for both digital and analog approaches;

– section 4.3 presents the digital microcontroller approach, the detection techniques and the communication protocol;

– section 4.4 presents the analog approach, with two UWB backscatterers used to modulate the tag mode;

– finally, section 4.5 discusses both systems, compares them with other state-of-the-art systems and draws the conclusions.

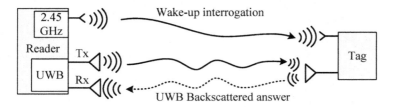

Figure 4.1. *Scheme of the hybrid UWB-2.4 GHz RFID system*

4.2. Wake-up system

This section explains the wake-up system, as well as the modulation schemes, common to both the microcontroller-based and the analog approaches. Interferences and coexistence with other commercial narrowband systems within the 2.4 GHz ISM band are also studied here. The wake-up system can be extended to other frequency bands by adjusting the antennas and matching networks.

4.2.1. Overview

Figure 4.2(a) shows a scheme of the wake-up system. It is composed of a 2.45 GHz antenna, followed by a matching network that adapts the impedance of the antenna to the impedance of a Schottky diode rectifier. The Schottky diode rectifier converts the RF signal to a DC signal. Finally, a Resistive-Capacitive (RC) block (with a parallel resistor and capacitor) is connected to obtain a DC signal. Figures 4.2(b) and (c) show the layout of the 2.45 GHz wake-up antennas used, consisting of a monopole and a dipole, fabricated on Rogers RO4003C (see Table 2.3). The antennas have 50 Ω impedance for an easy interchange. Figure 4.2(d) shows the simulated $|S_{11}|$ parameter (with Agilent Momentum) for both antennas, presenting a maximum gain of 1.77 dB and 2.69 dB for the monopole and the dipole, respectively.

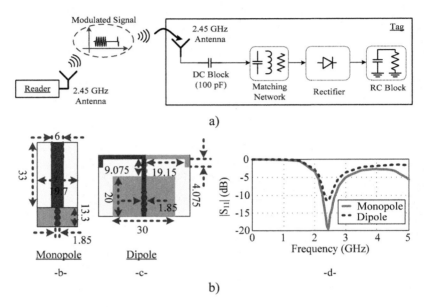

Figure 4.2. *a) Scheme of the wake-up system. Layouts (with dimensions in millimeters) of the wake-up b) monopole and c) dipole antennas at 2.45 GHz. d) Simulated |S$_{11}$| parameter of the monopole and dipole antennas*

4.2.2. Schottky diode-based detector

Diodes as RF detectors have studied for several years [COW 66] and are being used extensively in self-autonomous devices for energy harvesting and rectifying antennas (rectennas) [HER 14, JAB 10, GEO 10]. Detectors based on charge pump topologies are often used for RFID and rectennas applications [FER 11]. The simplest is using a series-connected diode structure. This structure has a lower impedance than a single diode structure, and the degradation on the efficiency is small compared with other structures with more diodes. Increasing the number of diodes increases the detected DC voltage, but due to losses of the diodes, the efficiency is decreased for low-power input levels.

The Avago HSMS-2852 series-connected Schottky diodes [AVA 14] are used as the core element of the envelope detector. At the 2.4–2.5 GHz band, the input impedance of the detector is mainly capacitive. There are several approaches to designing the matching network for the diode. For example, Figure 4.3 shows some options. The dimensions given are calculated for a Rogers RO4003C substrate at 2.45 GHz (see Table 2.3). In all cases, an RC block with R = 820 kΩ and C = 1 pF is considered. The simplest, least expensive approach is using a series transmission line with parallel open-ended stub (Figure 4.3(a)). The length of the lines is tuned to match the 2.45 GHz antenna impedance (normally 50 Ω) to the diode impedance. Another approach is to use a series transmission line, connected to a shunt-connected tuning capacitor (Figure 4.3(b)). The final approach considered is to use a series inductor and a short inductive transmission line, and a tuning shunt-connected capacitor (Figure 4.3(c)). The short inductive transmission line is needed to model the pad connection to solder the diode. It is also calculated so the inductor value is a commercial value. The matching networks have been adjusted for 50 Ω (for an easy interchange of wake-up antennas) and a low input power of –30 dBm, where a mismatch is more critical due to the low level of the detected DC voltage.

There are several advantages and disadvantages to each approach. When no discrete passive elements are used (Figure 4.3(a)), the realization of the tag is easier in a large-scale process (less elements are required to solder). In addition, it is easier to achieve a good impedance matching if a precise fabrication method is used for the tag layout. However, this approach also requires more layout space, resulting in a larger tag size. Therefore, this approach is preferred when the frequency of the wake-up is higher. On the other hand, with a matching based on lumped elements (Figure 4.3(c)), the resulting matching network is smaller, reducing the tag size. In this case, a precise mounting and soldering process of the passive elements onto the tag is critical in order to have a good impedance matching. Also, the passive elements (in the order of a few pF for the capacitors and nH for the inductors) must have very small tolerances. This is more difficult than simply adjusting the size of the transmission lines. The design shown in Figure 4.3(c) is an intermediate approach. Its tuning accuracy relies on the tolerance of the capacitor, but less space is required than in the approach shown in Figure 4.3(a).

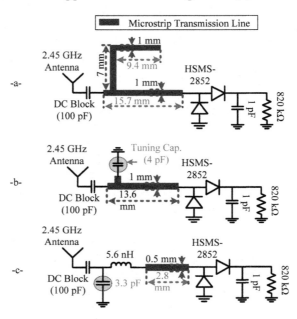

Figure 4.3. *Approaches for the matching network; a) series transmission line with a parallel open-ended stub; b) series transmission line with shunt-connected tuning capacitor; c) series inductor with short inductive transmission line and shunt-connected tuning capacitor. In red, elements tuned to achieve a matching at the desired frequency. For a color version of this figure, see www.iste.co.uk/ramos/rfid.zip*

In order to study the sensitivity, the diode behavior is simulated using the Agilent Advanced Design System (ADS) software. The structure shown in Figure 4.3(b) is chosen; however, it is important to note that the other structures present very similar results. The 2.45 GHz antenna is replaced by an S parameter port (50 Ω). Figure 4.4(a) shows a scheme of the simulation in ADS. The impedance Z_{in} is obtained by dividing the input voltage V_{in} by the input current I (monitored with a current probe), i.e. $Z_{in} = V_{in}/I$. Then, the input reflection coefficient is calculated as $\rho_{in} = (Z_{in} - 50)/(Z_{in} + 50)$. A harmonic balance (third order) simulation is performed, sweeping both the tone frequency and the input power. Figure 4.4(b) shows the simulated input reflection coefficient as a function of frequency, for input powers of −30 and −15 dBm. Figure 4.4(c) shows the impedance of the input reflection coefficient as a function of the input power at 2.45 GHz. The diode is matched (50 Ω) to the worst case scenario, i.e. the minimum input power. Figure 4.4(d) shows the output DC voltage as a function of the frequency for input powers of −30 and −15 dBm. As can be observed in Figures 4.4(b) and (d), even though the diode is not matched for higher input powers, the output voltage is still higher due to the power increase.

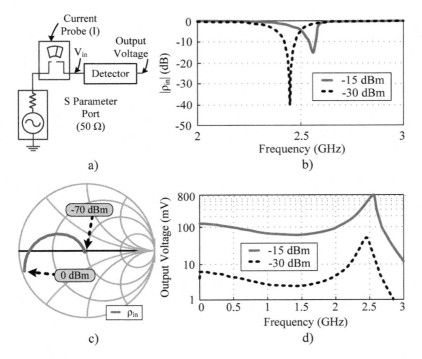

Figure 4.4. *a) Scheme of the diode detector simulation in ADS, b) simulated input reflection coefficient of the detector as a function of frequency for input powers of −30 dBm and −15 dBm, c) simulated impedance of the input reflection coefficient of the detector as a function of the input power at 2.45 GHz, d) simulated output voltage of the detector as a function of frequency, for input powers of −30 and −15 dBm*

Figure 4.5 shows the simulated and measured voltage at the output of the detector as a function of the RF power at the input. A detector prototype has been manufactured. A signal generator (RS SMA100A) is connected to the input, and a multimeter (Agilent 34410A) is connected to the output in order to measure voltage. The discrepancies at high input power are due to the internal model used in the simulator, which does not accurately take into account the breakdown effects in the diode current. The detected voltage is limited by the diode noise for low input power. Low input powers as small as –65 dBm can be detected according to the tangential sensitivity reported in the literature for similar zero bias devices [HEW 82]. A very good agreement between the simulated and the measured results can be observed. Theoretically, the reader would be able to wake-up the tag at a distance of 12.6 m in free space. This can be calculated considering an effective isotropic radiated power of +20 dBm output power, a wake-up antenna of 0 dBi in the tag and a +3 mV threshold. The 3 mV threshold has been chosen because it is a typical noise supply level in digital lines.

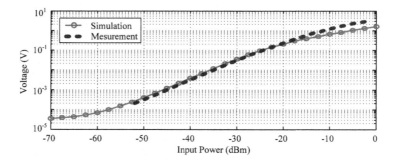

Figure 4.5. *Simulated (solid, red-circled line) and measured (solid blue line) wake-up circuit output voltage as a function of the input power. For a color version of the figure, see www.iste.co.uk/ramos/rfid.zip*

4.2.3. Reader: modulation schemes

The wake-up part in the reader is realized by means of a voltage-controlled oscillator (VCO). Specifically, a 2.45 GHz VCO (Mini-Circuits ROS 2793-119+). This oscillator sends a single-tone through a 2.45 GHz antenna.

An on–off keying (OOK) modulation is obtained by turning on/off the VCO power supply. A Mini-Circuits GAL-84+ medium power amplifier is connected at the output of the VCO. The transmitted power is +20 dBm and a monopole antenna is used in the reader. The PIC 16F1827 [MIC 14] microcontroller is used to interface with the PC by USB.

Figure 4.6(a) shows the scheme of the wake-up modulator, and Figure 4.6(b) shows a photograph of the interface board between the control PC and the oscillator. The interface board uses an FTDI FT232RL USB to serial converter circuit. Then, the PIC16F1827 is connected with the FT232RL (and hence the PC) using a standard serial interface.

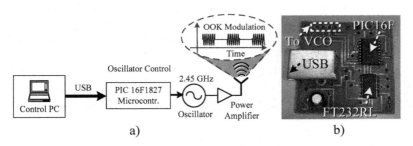

Figure 4.6. *a) Scheme of the wake-up modulator at the reader, b) photograph of the interface board between the control PC and the oscillator*

4.2.4. *Interferences and coexistence with other systems*

The ISM [LOY 14] band of 2.4–2.5 GHz is very crowded by home or office devices such as Wi-Fi and bluetooth, baby monitors or cordless phones. Therefore, it is important to study the potential influence of these devices on the detector circuit. A worst-case scenario is studied next to prove the robustness of the system. Figure 4.7(a) shows the measured spectrum of the wake-up signal with two continuous wave interferers placed at ±500 kHz, ±1 MHz and ±2 MHz, respectively. The distance between the reader and the tag is 40 cm. The spectrum has been measured with an R&S FSP Spectrum Analyzer. Figure 4.7(b) shows the corresponding time-domain signals at the output of the wake-up detector. The interferers are being transmitted with the same power as the wake-up signal (carrier-to-interference ratio of 0 dB). It can be observed that the wake-up signal is affected by the interferers, modifying its shape and mean level. Since a mean level estimator is used at the output of the rectifier, the pulse shape can be recovered by comparing the voltage to the mean level, as it will be explained in detail in section 4.3.2. Interferers closer than ±500 kHz of the carrier central frequency could alter the wake-up detection, as the signal is already distorted. Even though this could be a possible issue, in a real scenario this is not a problem. Wi-Fi and bluetooth networks are based on frequency hopping algorithms that do not emit continuously in a single frequency. Therefore, the wake-up signal would not be affected.

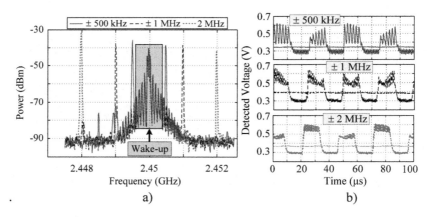

Figure 4.7. *a) Measured spectrum of the wake-up signal with several interferers, and b) corresponding time-domain signal at the output of the detector*

4.3. Microcontroller-based semi-passive UWB RFID system

4.3.1. *Introduction*

In this case, the core circuitry will be a commercial, ultra-low power microcontroller from Microchip Technologies, the PIC 16F1827 [MIC 14], already introduced in section 4.2.3 as the PC-VCO interface. The tag responds by backscattering a digital sequence. The use of a microcontroller permits the possibility of connecting both digital and analog commercial sensors, as well as the implementation of advanced communication schemes. Table 4.1 shows three of the most common low-power microcontrollers from Microchip [MIC 14], Texas Instruments [TEX 12] and Atmel [ATM 12]. As it can be observed, their features are very similar. The PIC16F1827 (from now on addressed as "the microcontroller") family has been chosen for its lower power consumption. The microcontroller integrates the comparator used with a diode-based detector to wake-up the tag (see section 4.2). The microcontroller also stores the ID code, acquires data from the sensors and modulates the radar cross-section (RCS) of the antenna changing the biasing of a PIN diode. In order to save battery, the microcontroller in the tag rests in a low-power consumption mode (sleep mode) until it is interrogated by the reader using the wake-up link. The current consumption in sleep mode is under 100 nA.

Feature	PIC16F1827 [MIC 14]	MSP430FR5969 [TEX 12]	ATtiny43U [ATM 12]
Internal clock speeds	31 KHz to 32 MHz	10 KHz to 24 MHz	128 KHz to 8 MHz
Supply voltage	1.8–5.5 V	1.8–3.6 V	0.7–5.5 V
Supply current (Operation @ 1 MHz)	75 μA	100 μA	400 μA
Minimum supply current (sleep)	30 nA	40 nA	150 nA
ADC converter resolution	10 bit	12 bit	10 bit
Digital bus compatibility	Yes, I^2C	Yes, I^2C	Yes, I^2C

Table 4.1. *Comparison between commercial low-power microcontrollers*

Figure 4.8 shows an image of the tag. It consists of a monopole UWB tag (see section 2.6.2) loaded with a PIN diode, and the 2.45 GHz detector shown in Figure 4.3(c) connected to the 2.45 GHz monopole antenna presented in section 4.2.1. Both are connected to the microcontroller, which is powered using a 3 V lithium battery. The tag size is 70.93 mm × 50.97 mm and it is fabricated on a Rogers RO4003C substrate (see Table 2.3).

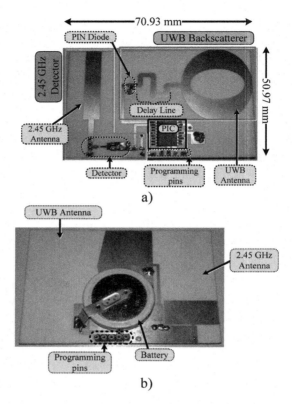

Figure 4.8. *Photograph of the tag: a) top face and b) bottom face*

4.3.2. *Microcontroller: tag core logic*

To minimize the tag size, the rectifier is realized using the approach of mainly passive elements (see section 4.2.2), and its circuit is shown in Figure 4.9(a). The rectifier output is connected to the negative input of the microcontroller ("data") internal comparator, as shown in Figure 4.9(b). The threshold voltage for this comparator is generated by a slicing RC circuit ($R = 100$ kΩ, $C = 3$ nF) at the output of the detector. In this way, the threshold automatically adjusts to the mean value of the detected signal. The comparator generates an interrupt when the wake-up signal is greater than the threshold voltage, initiating the microcontroller program. Using an internal comparator increases the tag battery lifetime, since it avoids additional circuitry that would increase the current consumption of the tag during sleep periods.

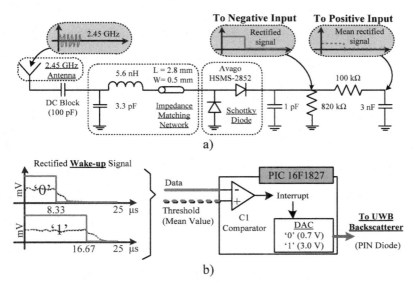

Figure 4.9. *a) Scheme of the 2.45 GHz wake-up detector circuit with microcontroller, b) scheme of the wake-up signal for "0" and "1" cases and their detection using the microcontroller internal comparator*

In order to send information from the reader to the tag using the wake-up signal, a pulse-width modulation (PWM) scheme is used, as it is also shown in Figure 4.9(b). Depending on the high level time of the rectified signal, a "0" or a "1" bit is considered. A 33% of the period at a high level codes a "0", and a 66% codes a "1". In this prototype, the period of the PWM signal for each bit is set to 25 μs, meaning a speed of 40 kbps. This speed is enough for the amount of information that a RFID reader needs to send (mainly interrogation requests and configuration commands). The microcontroller program clock frequency has been set to a reduced value (1 MHz) to minimize the tag power consumption.

One of the main advantages of this topology is that the reader controls the tag clock using the wake-up link, and there is not a pulse generator in the tag. Therefore, there is no need for a synchronization method between the reader and the tag clocks, as performed in [BAG 09].

4.3.3. *UWB backscatterer design and evaluation*

The backscatterer is composed of a UWB antenna connected to a delay line of length L and characteristic impedance Z_c. As discussed in section 2.2, a two-port circuit model is used to explain the time-domain signal scattered by a UWB antenna connected to a transmission line. Here, the delay line is loaded with a PIN diode (NXP Semiconductors BAP51-03) [NXP 04a] biased through a 33 kΩ resistor. It is shown in Figure 4.10(a). The bias of the PIN diode is controlled by the microcontroller, implementing a logical "1" when the diode is ON and a logical "0" when it is OFF.

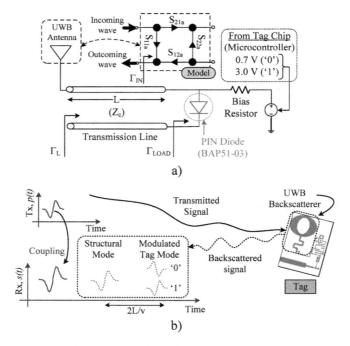

Figure 4.10. *a) Model of the modulated backscattering UWB PIN-loaded antenna, b) scheme of the transmitted and backscattered UWB signals*

Figure 4.10(b) shows a scheme of the transmitted (Tx) and received (Rx) signals at the reader. The transmission line connected to the tag's antenna (see section 2.2) is now ended with a PIN diode. The microcontroller sends a "0" or "1" signal by setting the output voltage to a low (0.7 V, instead of 0 V, will be explained next) or high (3 V) level, respectively. In this way, the PIN diode is biased from an open circuit ($\Gamma_{LOAD}= 1$) to a short circuit ($\Gamma_{LOAD} = -1$), depending on the current that flows through it. Therefore, the "0" and "1" states will have tag modes with opposite phases.

In order to study the time-domain response of the BAP51-03 PIN diode, a characterization board is used. It is shown in Figure 4.11(a). It consists of a microstrip transmission line fabricated on Rogers 4003C substrate. The transmission line is terminated with the PIN diode. Some extra pads are also added to solder a polarization resistor and the power supply wires to polarize the diode. The reflection coefficient is measured between 0.1 and 10 GHz with 1601 points using the Agilent E8364C network analyzer (VNA). In order to calibrate the VNA at the same plane where the PIN diode is soldered, a custom calibration kit (also shown in Figure 4.11(a)) is used to perform an openshortload calibration [AGI 04]. A 33 kΩ 0603 surface mount (SMD) resistor is used to bias the diode, acting as a broadband RF block. An Agilent 34410A multimeter is connected in series with a programmable DC power source for current monitoring. Figure 4.11(b) shows the diode resistance as a function of the forward current from the diode datasheet. According to the diode datasheet, a current greater than 500 µA would be necessary to achieve a short-circuit state. Figure 4.11(c) shows the time-domain response (after applying the inverse Fourier transform to the VNA acquired signal) of the calibrated S_{11} parameter of the diode for several diode forward currents I_F. It can be seen that the pulse is inverted (meaning a ±180° change in phase) for forward currents equal or greater than 30 µA. This is due to the change in the reflection coefficient introduced by the diode, from an open circuit to a short circuit. It can be observed that when biasing the diode with just 30 µA, the phase change is perfectly detected. Moreover, for forward currents greater than 130 µA the amplitude values start to saturate, since the reflection coefficients are similar. Therefore, it is proved that this diode can be used to change the reflection coefficient of a load at UWB frequencies, and a current of 500 μA is not required.

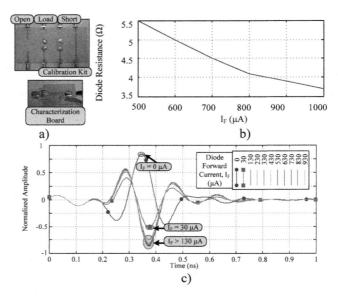

a)

b)

c)

Figure 4.11. *a) Characterization board and custom calibration kit for the PIN diode, b) diode resistance as a function of the forward current, according to the manufacturer's datasheet, c) time-domain response of the S_{11} parameter of the diode for forward currents from 0 to 930 μA. For a color version of the figure, see www.iste.co.uk/ramos/rfid.zip*

Using the time-domain radar, the tag is read at a fixed 50 cm distance for different diode forward currents. The background subtraction technique and the time-windowing technique from section 2.4.1 are applied to the received signal. Figure 4.12(a) shows the tag's time-domain response for PIN diode currents from 0 to 730 μA. The time delay of the structural modes for all currents remains the same, as expected. The tag modes present an inverted phase when the forward current is greater than 0 μA. There is a small difference between 30 μA and >130 μA states due to the change in the resistance when the forward current increases. For currents greater than 130 μA, the difference is not noticeable. A forward current of 70 μA is chosen for the "1" state. This is to achieve a compromise between consumption and performance.

In all cases as shown in Figure 4.12(a), it can be seen that the amplitude of the tag mode also changes when biasing the diode. When the diode current increases, the diode resistance decreases. Then, the magnitude of the reflection coefficient increases, resulting in an increase in the tag mode amplitude. Ideally, the maximum would correspond to a short circuit. Figure 4.12(b) shows the measured tag-to-structural mode ratio for currents from 0 to 70 µA. It corresponds to the tag mode peak amplitude over the structural mode peak amplitude. We can see that the maximum amplitude difference for two diode currents is achieved with a forward current of 5 µA for the "0" state. Therefore, the "0" state is generated by biasing the diode with 0.7 V with the 33 kΩ bias resistor (see Figure 4.9). The state change can therefore be detected as an amplitude change rather than as a phase change, permitting the use of the wavelet processing technique (see section 2.4.2). It will be addressed in detail in section 4.3.7.

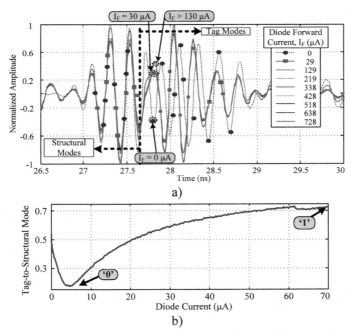

Figure 4.12. *Time-domain response of the UWB backscatterer for a) PIN diode forward currents from 0 to 930 µA and b) tag-to-structural mode ratio for currents from 0 to 70 µA. For a color version of the figure, see* www.iste.co.uk/ramos/rfid.zip

4.3.4. *Differential coding and detection techniques*

Instead of subtracting the same empty room background scenario from all signals, an alternative technique is used for the digital tag. As explained in section 2.4.1, the background subtraction eliminates clutter assuming that it does not change in time. In addition, it is assumed that the structural and tag modes have a finite time duration and the delay between the two signals introduced by the delay line is enough to separate them. However, from the previous results discussed in Chapters 2 and 3, a residual interference between the two modes can be observed. In order to reduce this interference, the length of the delay line should be increased, resulting in a large tag size and a reduction of tag mode amplitude due to the transmission line losses. To solve this problem, a differential coding schema has been adopted. For each measured RAW signal $s_i(t)$, the differential signal $d_{i-1}(t)$ is written as:

$$d_{i-1}(t) = s_i(t) - s_{i-1}(t),$$ [4.1]

where $i = 2,3,...,n$ (n is the number of bits in a frame) and $s_{i-1}(t)$ is the preceding RAW signal corresponding to the measured previous state. In this manner, the difference between one state and its preceding state is obtained. From the chipless theoretical expressions in section 2.2, it can be seen that the structural mode will be cancelled because it does not change with the load or state. The clutter is also considerably reduced, assuming that it does not change between one bit and another. For two consecutive identical states, i.e. two "0" or two "1" states, all the differential signals have an amplitude near zero. This is because $s_i(t)$ and $s_{i-1}(t)$ are very similar both on the structural and tag modes. On the contrary, for one "0" state followed by a "1" state or viceversa, the differential signal has no structural mode, and the tag mode is greatly increased. This is due to the fact that $s_i(t)$ and $s_{i-1}(t)$ have identical structural modes, but tag modes with different amplitudes and phases, as shown in Figure 4.12(a).

This differential coding technique also permits measurement of the tag without the need for measuring the empty room response. Moreover, since the structural mode is removed, there is no need for a small structural mode in order to increase the structural tag mode ratio, as was performed in [HU 08]. It is similar to a differential RCS approach used in passive UHF RFID [NIK 07]. Taking into account expression [2.5], and considering Γ_{L0} and Γ_{L1} the reflection coefficients at the load for states "0", and "1", respectively:

$$\Gamma_{in0} = S_{11a} + S_{21a}S_{12a}\Gamma_{L0},$$ [4.2]

$$\Gamma_{in1} = S_{11a} + S_{21a}S_{12a}\Gamma_{L1},$$ [4.3]

and the differential reflection coefficient at the input of the tag can be calculated as:

$$\Delta\Gamma_{in} = \left|\Gamma_{in0} - \Gamma_{in1}\right| = S_{21a}S_{12a}\left|\Gamma_{L0} - \Gamma_{L1}\right|,$$ [4.4]

where S_{12a} and S_{21a} can be interpreted as the tag's antenna gain. The differential signal has a tag mode. Therefore, the CWT technique can be applied to improve the signal-to-noise ratio, as performed in section 2.4.1. To distinguish whether there is or not a state change, a threshold must be considered to compare with the maximum peak amplitudes. Theoretically, a fixed value of 0.5 should be considered for the threshold. The "non-changed" states have an amplitude near 0, and the "changed" states have an amplitude near 1. However, in practice, this threshold must be variable. Random noise and propagation can produce differences between two identical bits. For an 8-bit sequence, the CWT of the differential signal is calculated, and then the maximum peaks are obtained. These peaks are classified as "non-changed" or "changed" with a threshold of 0.5. The variable threshold Th is then calculated as:

$$Th = \frac{1}{2}\left(\overline{NonChg} + \overline{Chg}\right),$$ [4.5]

where $\overline{NonChg}$ and $\overline{Chg}$ are the mean values of the "non-changed" and "changed" states, respectively, for a frame. Figure 4.13 shows a constellation diagram for the "non-changed" and "changed" states. It can be interpreted as an amplitude shift keying modulation with a variable threshold.

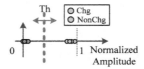

Figure 4.13. *Constellation diagram for the "changed"
and "non-changed" states at the tag after differential signal.
For a color version of the figure, see www.iste.co.uk/ramos/rfid.zip*

4.3.5. Communication protocol

The communication protocol between reader and sensor consists of four main steps, as shown in Figure 4.14:

1) the reader sends the tag ID byte and wakes up the tag;

2) the reader sends mode configuration;

3) the reader sends bit requests and reads the tag answer;

4) the reader sends last acknowledgment, closes the transmission and the tag goes into sleep mode.

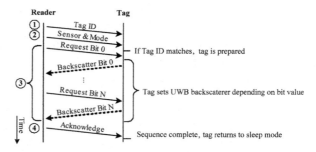

Figure 4.14. *Communication protocol between the tag and the reader*

Step 1 is essential in order to avoid false wake-ups, either due to signals coming from other systems or from any non-authorized party, which would discharge battery. Step 2 permits information to be sent to the sensor of the operation mode. For instance, it can request the instantaneous measurement of one or several sensors (mode 1). It can program a connected sensor to perform a series of measurements autonomously and store them (mode 2). Finally, it can ask a sensor to download the stored measurements (mode 3). Figure 4.15 shows, as an example, the measured control voltages at the reader and at the tag associated with the steps of this protocol (in this communication 10 bits are transmitted).

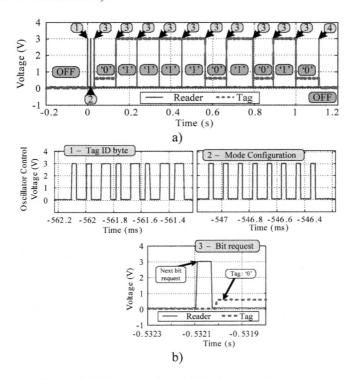

Figure 4.15. *Measured control voltages at the reader and tag; a) full sequence, b) zoomed steps 1, 2 and 3. For a color version of the figure, see www.iste.co.uk/ramos/rfid.zip*

4.3.6. *System scalability, applications and sensor integration*

Since a programmable microcontroller is used, more sophisticated anti-collision protocols or data encryption algorithms could be implemented [MAG 09]. The main limitation is the computational power of the microcontroller. In this case, the PIC microcontroller (like other microcontrollers, as presented in Table 4.1) has selectable internal clock frequencies. In the case of the PIC, from 31 KHz to 32 MHz, higher clock frequencies permit the implementation of faster and more complex cryptographic algorithms, at the expense of a larger battery consumption.

The bit sequences transmitted to (wake-up) and received from (backscatter) the tag are variable, and can be changed depending on the application. For instance, an application where the tag has only to respond a sensor read requires few bits. However, an identification application with cryptography requires a significant amount of information. The advantage of using a programmable microcontroller is that it can be modified depending on the requirements.

Several sensors can be integrated in the RFID tag, connected to the microcontroller either by means of an analog-to-digital converter (ADC) or by using an I^2C bus, as in [DE 13]. Figure 4.16 shows a scheme of the integration of several sensors and how are they read. Analog sensors are read using the microcontroller's internal ADC. Depending on the specifications, analog sensors may require a signal conditioning circuit to adapt the sensor output to a voltage between 0 and 3 V, which is the normal margin when the microcontroller is powered by a 3 V lithium battery. Digital devices or sensors are connected through the I^2C bidirectional bus. The signal conditioning and acquisition is performed by the sensor itself, and is determined by the manufacturer.

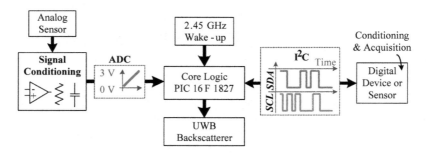

Figure 4.16. *Integration of sensors in the digital microcontroller-based tag*

4.3.7. *Results*

To verify the system, the tag sends the sequence "00101011" when interrogated by the reader. The sequence is answered by backscattering: the tag microcontroller biases the PIN diode from 5 µA to 70 µA. In order to synchronize the tag and the reader, the tag only changes its state (i.e. sends the next bit of the sequence) when the reader sends a wake-up signal, meaning that the reader is ready for the next bit. Figures 4.17(a) and (b) show the time-windowed UWB RAW signals for all states obtained with the radar setup at 50 cm. All signals are normalized with respect to all of their maximum absolute amplitude. A difference between "0" and "1" states can be perfectly observed. Figures 4.17(c)–(d) show the same signals after applying the background subtraction technique (see section 2.4). It can be seen that the noise has been reduced, and the difference between "0" and "1" states is more noticeable.

Since a difference in the signal amplitude between both states is also noticeable, the CWT is applied to all signals independently. The maximum magnitude cuts of the CWTs are shown in Figure 4.18(a). It can be observed that the tag modes for "1" states have a greater amplitude than "0" states. Finally, Figure 4.18(b) shows an image of the CWT signals sequence. It can be observed that the states are correctly detected, and the bit sequence can be obtained by the reader.

The benefits of the differential signal approach explained in section 4.3.4 are now presented. Figure 4.19 shows the differential signals for the same "00101011" sequence. It can be seen that the structural modes are removed by the subtraction of the previous signal for all cases, but the tag modes appear when there is a state change (bit change).

Figure 4.20(a) shows the CWT of the differential signals for the same sequence ("00101011"). All signals are normalized with respect to all of their absolute maximum. Since the measurement at 50 cm is very close to the reader, the dynamic threshold is 0.5016, very close to the theoretical value of 0.5. Figure 4.20(b) shows the image sequence of the differential signals at 50 cm. The state changes occur according to the sequence the tag is responding.

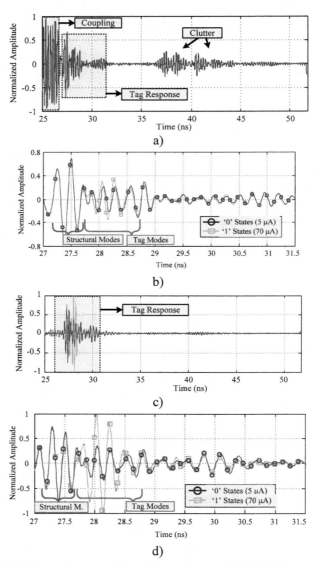

Figure 4.17. *Time-windowed signals for sequence "00101011" answered by the tag at 50 cm; a) and b) RAW, and c) and d) after background subtraction. For a color version of the figure, see www.iste.co.uk/ramos/rfid.zip*

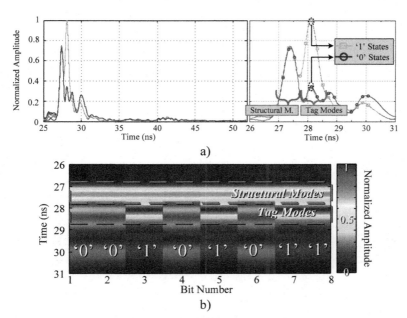

Figure 4.18. *a) Maximum magnitude cuts of the continuous wavelet transforms of the background-subtracted signals at 50 cm, b) image sequence of the CWT maximum magnitude cuts of the background-subtracted signals at 50 cm. For a color version of the figure, see www.iste.co.uk/ramos/rfid.zip*

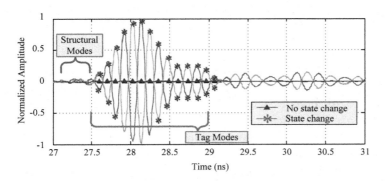

Figure 4.19. *Differential signals for the bit sequence "00101011" answered by the tag. For a color version of the figure, see www.iste.co.uk/ramos/rfid.zip*

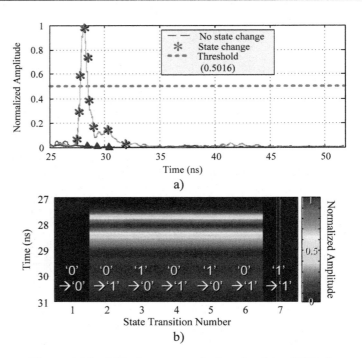

Figure 4.20. *a) Maximum magnitude cuts of the CWT of the differential signals at 50 cm, b) image sequence of the CWT maximum magnitude cuts of the differential signals at 50 cm. For a color version of the figure, see www.iste.co.uk/ramos/rfid.zip*

Figure 4.21(a) shows the same sequence read at 8.5 m. The time window has been shifted in order to detect the tag at distances longer than 50 cm. Now the noise is higher, meaning that the amplitude difference between a state change case and a no state change case is smaller. The dynamic threshold in this case is 0.5750. Figures 4.21(b) and (c) show the image sequence of the differential signals at 8.5 m. Similarly to the 50 cm case, the state changes occur according to the sequence. In this last case, however, the differential tag modes are not perfectly aligned due to the higher noise at 8.5 m. However, the reader can still clearly detect whether there is or not a state change.

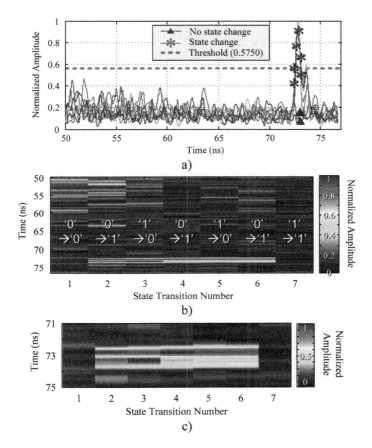

Figure 4.21. *a) Maximum magnitude cuts of the CWT of the differential signals at 8.5 m, b) image sequence of the CWT maximum magnitude cuts of the differential signals at 8.5 m and c) zoomed image of the sequence. For a color version of the figure, see www.iste.co.uk/ramos/rfid.zip*

4.4. Analog semi-passive UWB RFID system

4.4.1. *Introduction*

Here, the core circuitry is composed of simple conditioning circuits. The sensor itself directly modulates the tag's RF backscattered response. This response is, instead of a digital sequence, an analog value related with the physical magnitude.

The system works as shown in the flow diagram of Figure 4.22(a), using the same hybrid 2.45 GHz – UWB system presented in section 4.1.2. First, the reader measures the background scene with the UWB radar, which consists of the tag being deactivated. Next, the reader sends a 2.45 GHz calibration signal with a "calibration" state. The tag sets itself to a known calibration state, which is independent of the sensor value. Then, the reader collects the backscattered calibration answer via UWB. After, the reader sends another calibration signal with a "sensor" state. The tag sets itself to a state which, in this case, depends on the sensor, and the reader collects the backscattered answer via UWB.

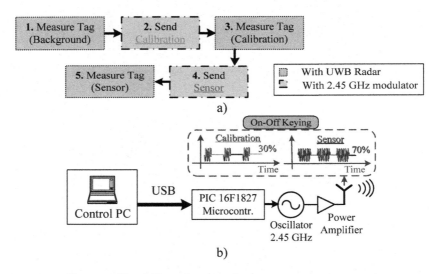

Figure 4.22. *a) Flow diagram of one sensor measurement, b) scheme of the signals sent from the reader for the analog system. For a color version of the figure, see www.iste.co.uk/ramos/rfid.zip*

Similarly to section 4.3, the reader uses an OOK modulation over the 2.45 GHz signal in order to generate the states in the tag. The tag itself integrates a wake-up detector, as explained in section 4.2. The two calibration signals consist of a "calibration" state, with a duty cycle of the 30% of the OOK, and a "sensor" state, with a duty cycle of 70% of the OOK. They are shown in Figure 4.22(b). These signals are not interpreted as digital bits by the tag. They activate analog circuitry composed of operational amplifiers and DC switches, as will be explained in detail in section 4.4.4.

4.4.2. *Switch-based UWB backscatterer*

Figure 4.22 shows a diagram of the sensor tag, based on a UWB backscatterer topology, and the signals sent from (TX) and backscattered to the reader (RX). The tag is based on a UWB antenna connected to a delay line (L_1), which is in turn connected to a commercial RFIC single-pole double throw (SPDT) switch from Skyworks (AS186-302LF) [SKY 07]. The two outputs of the switch are connected to two delay lines, L_2 and L_3, respectively. For identification purposes, it is desirable for the load to be an open or short circuit in order to maximize the amplitude of the tag mode. The tag identification information (ID) is coded in the time delay between the structural and tag modes, which depends on transmission line length L_1. Here, however, transmission line L_1 is connected to input J_1 of the SPDT switch. Output J_2 of the switch is connected to an open-ended delay line of length L_2, whereas output J_3 is connected to an open-ended delay line of length L_3. It is important to note that length $L_2 \neq L_3$. Hence, the tag modes associated with positions J_2 and J_3 of the switch have two different structural-to-tag mode delays, and so each can be distinguished from the other.

Switch actuation voltage V_1 is controlled by the sensor by means of a conditioning circuit. Hence, the switch insertion loss (and therefore the tag mode amplitude) depends on the sensor value. According to the manufacturer, the switch is operated as follows. To connect J_1 with J_2, the V_1 input must be in a high voltage state (3 V), whereas the V_2 is in a low voltage state (0 V). Similarly, to connect J_1 with J_3, $V_1 = 0$ V and $V_2 = 3$ V. Here, instead of operating the switch with 0 V or 3 V, intermediate values of V_1 within this range are considered to intentionally change its insertion loss and modulate the amplitude of the tag mode associated with the J_2 output (which depends on V_1).

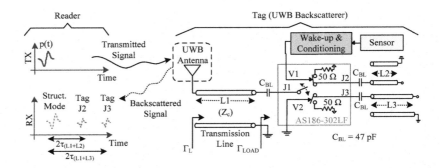

Figure 4.23. *Diagram of the switch-based backscatterer tag and the TX and RX signals at the reader. For a color version of the figure, see www.iste.co.uk/ramos/rfid.zip*

Figure 4.24 shows the $|S_{21}|$ scattering parameter considering J_1 as port 1 and J_2 as port 2, measured by the Agilent E8364C VNA (see section 2.3.1). A photograph of the board with the switch is also shown. The frequency is swept from 0.1 to 12 GHz with 1601 points at +0 dBm of output power. The V_1 pin of the switch is driven from 0 to 3 V, while the V_2 pin is at a low (0 V) state for all measurements. Evidently, the amplitude of the scattering parameter can be modulated by changing V_1. For the case of 0 V, the transmission is minimal as expected. For the case of 3 V, however, the transmission is maximal at the frequency band of the switch (0–6 GHz), with an insertion loss similar to the insertion loss provided by the manufacturer [SKY 07].

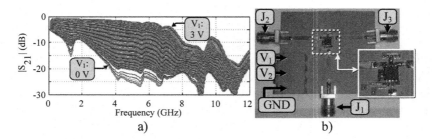

Figure 4.24. *a) $|S_{21}|$ between J_1 and J_2 as a function of V_1 voltage, b) photograph of the board with the switch. For a color version of the figure, see www.iste.co.uk/ramos/rfid.zip*

Figure 4.25(a) shows the time-domain response of the same S_{21} parameter, but in this case obtained by applying the absolute value to the inverse Fourier transform of S_{21}. The peaks are detected and shown in Figure 4.25(b) as a function of V_1. There is an exploitable zone in the 1–1.4 V region (marked with a shadow), which can be used to modulate the tag mode. Five measurements are overlapped. This behavior is very repeatable since the mean standard deviation is 7.85×10^{-5}. It is important to note that, as shown in Figure 4.25, all the measurements are normalized with respect to the maximum case, i.e. 3 V.

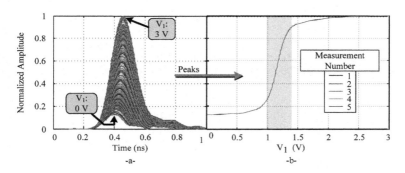

Figure 4.25. *a) Time-domain response of the S_{21} parameter as a function of V_1, b) detected peaks of the S_{21} time-domain response as a function of V_1. For a color version of the figure, see www.iste.co.uk/ramos/rfid.zip*

Next, an emulated tag that follows the scheme, as shown in Figure 4.23, is measured. The aim is to demonstrate that the modulation of the tag mode amplitude due to the switch can be detected remotely. The tag is composed of a UWB antenna connected to a coaxial line of length L_1, with a round-trip delay of 2.3 ns. The other end of L_1 is connected to input J_1 of the switch. Outputs J_2 and J_3 of the switch are left to an open circuit without any line ($L_2 = 0$) and connected to another coaxial line of length $L_3 = L_1$, respectively. Tag modes 1 and 2 are therefore separated with 2.3 ns between them. V_1 here is driven by a controllable power supply. The tag is measured using the time-domain radar (see section 2.3.2). Figure 4.26 shows the time-domain response of the tag as a function of V_1 measured at a 50 cm tag-reader distance. The CWT processing technique (see section 2.4.2) is applied to all the measurements to reduce noise. All measurements are normalized with respect to the absolute maximum, which is the structural mode. We can see that while the structural modes remain invariant, the tag modes vary depending on V_1.

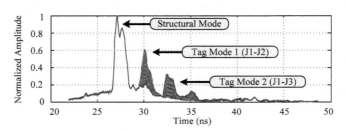

Figure 4.26. *Time-domain response of the tag as a function of V_1, measured by the UWB radar. For a color version of the figure, see www.iste.co.uk/ramos/rfid.zip*

Figure 4.27 shows the peaks of the normalized tag modes (1 and 2) for eight consecutive measurements with the radar. Repeatable behavior is again observed, now using the UWB radar, since mean standard deviations of 7.85×10^{-4} and 1.1×10^{-3} are obtained for tag modes 1 and 2 respectively. In this section, tag mode 1 is exploited to sense and tag mode 2 is used for identification by changing length L_3.

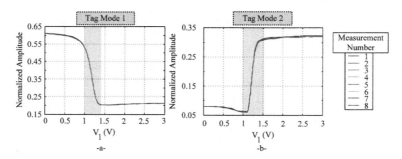

Figure 4.27. *a) Detected peaks of tag mode 1 (left) and b) tag mode 2 as a function of V_1*

Finally, Figure 4.28 shows the current consumed by the switch as a function of V_1 ($V_2 = 0$ V). We can see that the maximum current consumption is 0.25 µA, which makes it a perfect candidate for integration as an RF transducer in battery-powered wireless sensors.

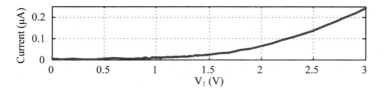

Figure 4.28. *Current drained by the AS-182-320LF switch as a function of V_1 ($V_2 = 0$ V)*

4.4.3. *PIN diode-based UWB backscatterer*

Figure 4.29 shows a scheme of the UWB backscatterer and the signals sent from the reader (TX) and backscattered at the tag (RX). The tag is composed of a UWB antenna connected to a delay line that is, in turn, loaded with a low-cost PIN diode (model NXP BAP64-03). It is similar to the backscatterer discussed in section 4.3.3, but now another diode with a different resistance curve (as a function of forward current) is used.

Here, the load Z_{LOAD} is modulated because the resistance of the PIN diode in forward polarization depends on the current that flows through it [NXP 04b]. This current is controlled by the wake-up and conditioning circuit, and in turn the sensor. Figure 4.30(a) shows the measured backscattered signal of a time-coded chipless tag loaded with the BAP64-03 PIN diode, biased with a 10 kΩ resistor, as shown in Figure 4.29. The measurement is performed using the time-domain radar (see section 2.3.2), at a tag-reader distance of 30 cm. The diode bias voltage (V_{BS}) is manually swept from 0 to 3.3 V (current between 0 and 260 µA). Each line represents a different current. In Figure 4.30(b), the CWT has been applied to the measured signal to improve the signal-to-noise ratio (see section 2.4.2). All signals are normalized with respect to their own maximum amplitude, which is the structural mode. It can be observed that the tag mode amplitudes vary depending on the current that flows through the diode.

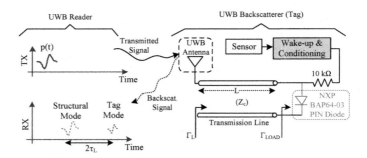

Figure 4.29. *Scheme of the sent and received signals between the reader and tag for the PIN diode-based analog backscatterer*

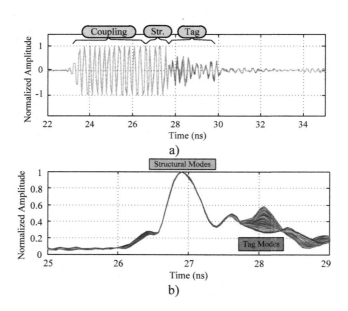

Figure 4.30. *Measured time-domain response of the analog PIN diode-based tag depending on the diode current; a) raw signal and b) signal after applying the CWT. For a color version of the figure, see www.iste.co.uk/ramos/rfid.zip*

Figure 4.31 shows the Structural-to-tag mode ratio as a function of the diode current, similarly as performed in section 4.3.3. The ratio goes from a large value at 0 μA, which is equal to an open circuit state ($Z_{\text{LOAD}} \to \infty$), to a low value at around 55 μA, which is equal to a matched load state ($Z_{LOAD} \rightarrow Z_C$). It is important to note that the ratio is not equal to zero at the minimum of the curve, mainly because of parasitic effects of the diode. The shaded area in Figure 4.31 shows a low-power consumption zone of the curve that can be exploited. Considering a 10 kΩ polarization resistor, this zone corresponds to a bias voltage range between 0.45 and 1.15 V, and current between 0 and 55 μA. Contrary to the BAP51-03 diode from section 4.3.3, the exploitable continuous zone here is larger for similar diode currents. The BAP64-03 is a diode used for limiter circuits, and therefore its resistance curve is more linear. Consequently, it is chosen for this specific design.

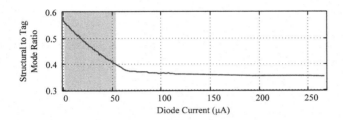

Figure 4.31. *Structural-to-tag mode ratio as a function of the diode forward current*

4.4.4. Detector circuit design

Next, the detector circuit implemented in the tag to detect both "calibration" and "sensor" signals introduced in section 4.4.1 is explained. This detector is based on the parallel open-ended stub design for the matching network (see section 4.2.2 and Figure 4.3(a)). As will be described in the sensor implementations in Chapter 5, the tag sizes are larger than the microcontroller-based approach, because of the wake-up and conditioning circuits. Therefore, the matching network is not required to be miniaturized: it does not imply a substantial increase in the overall tag size.

The detector is shown in Figure 4.32. It is fabricated on the same Rogers RO4003C substrate (see Table 2.3). It consists of a dipole antenna (see Figure 4.2(c)) connected to an Avago HSMS-2852 Schottky diode rectifier. A 100 pF DC-block capacitor followed by an

L-shaped matching network connects the antenna a with the detector. At the output of the Schottky diode, a 1 nF capacitor and an 820 kΩ resistor are shunt-connected to obtain the rectified voltage from the diode. A Texas Instruments TLV2401 low-power operational amplifier is used as a comparator, providing a 0 V output when the rectifier does not detect voltage and 3 V when it does. The comparator threshold is obtained by an RC estimator ($R = 82$ kΩ, $C = 1$ μF). Finally, the output of the comparator is connected to another RC estimator ($R = 820$ kΩ, $C = 1$ μF), which provides the mean value of the OOK signal between 0 and 3 V. This mean value depends on the duty cycle, but not on the tag-to-reader distance.

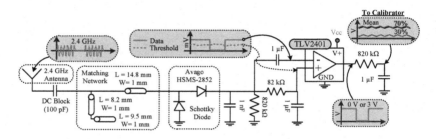

Figure 4.32. *Scheme of the 2.4 GHz detector circuit*

The circuit is shown in Figure 4.33(a). A time diagram of the main signals involved in the calibrator circuit is shown in Figure 4.33(b). The circuit consists of a (first) Texas Instruments TLV2402 dual channel operational amplifier, which acts as a comparator, detecting whether the mean value of the wake-up OOK signal exceeds two predefined voltage thresholds V_{th1} and V_{th2}. If the mean value exceeds V_{th1} but is below V_{th2} (meaning a 30% duty cycle), a second TLV2402 and a Maxim MAX4523 switch are powered from the first TLV2402 output 30P. This corresponds to the "calibration" state. In this case, the amplifier B2 of the second TLV2402 provides a V_{cal} voltage for the switch input COM2. This amplifier B2 operates as a buffer to render the V_{th2} threshold independent from V_{cal}. Since the mean value is not above V_{th2}, the switch will provide V_{cal} at its output ($V_{BS} = V_{cal}$). V_{cal} is a voltage that sets a stable, known state of the backscatterer. It can be generated using a simple resistive divisor circuit. V_{cal} is chosen as:

– PIN diode-based (BAP64-03) backscatterer: $V_{cal} = 1.33$ V;

– switch-based (AS186-302LF) backscatterer: $V_{cal} = 3$ V.

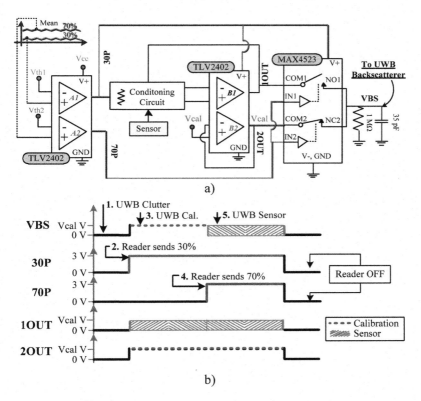

a)

b)

Figure 4.33. *a) Scheme of the calibrator circuit,*
b) time diagram of the signals

When the mean value exceeds V_{th2} (meaning a 70% duty cycle), the first TLV2402 outputs (30P and 70P) are at up state and the MAX4523 output (V_{BS}) will correspond to input COM1. This corresponds to the "sensor" state. The COM1 voltage depends on the sensor value. This COM1 sensor-dependent voltage is generated from the conditioning circuit. The conditioning circuit must adapt 1OUT (and hence V_{BS}) to the requirements of the backscatterer used (see sections 4.4.2 and 4.4.3):

- PIN diode-based (BAP64-03) backscatterer:
  1OUT = [0.45 , 1.15] V;

- switch-based (AS186-302LF) backscatterer:
  1OUT = [1 , 1.4] V.

Finally, the output of the MAX4523 switch is ended with a 1 MΩ resistor and a 35 pF capacitor due to the manufacturer's requirements.

Figure 4.34 shows the flow diagram of the signal processing carried out for all measurements. First, the background is measured without sending any wake-up signal. Then, the "calibration" state is measured with the wake-up duty cycle at 30%. Finally the "sensor" state is measured with the wake-up duty cycle at 70%. The background is subtracted from the measurements at "calibration" and "measurement" states. Then, the CWT is applied to both states independently. Finally the "measurement" is calibrated with respect to the "calibration", and the maximum peak of the tag mode is obtained.

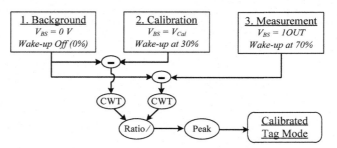

Figure 4.34. *Flow diagram of the signal processing with the analog semi-passive UWB RFID system*

4.5. Discussion, comparison between systems and conclusions

This chapter has presented the following two sensing platforms for remote identification and sensing based on time-coded UWB RFID:

- a digital approach, based on a commercial microcontroller;

– an analog approach, based on operational amplifiers, DC switches and passive elements.

The digital approach permits the advanced sensors, such as I^2C. It has a long read range (up to 8.5 m) and integration of is more robust than the analog approach because only binary data are transmitted wirelessly. In addition, it is scalable and can implement advanced applications such as cryptography, since it has a programmable microcontroller. Table 4.2 shows a comparison between the digital system and other state-of-the-art digital systems.

The analog approach provides a simpler alternative, where the sensor directly modifies the amplitude of the RF signal (tag mode). It is potentially less expensive if all the components were integrated. The reading is also faster: only three RF measurements are required, compared to a minimum of 10 for an ADC reading of the digital microcontroller. A calibration system has been used to independize the sensor read from the tag-reader distance or angle. Despite its advantages, the analog approach is more vulnerable to noise, and, therefore, its read range is more limited.

A comparison between the analog and digital approaches explained here will be performed in Chapter 5 based on actual sensor implementations of both systems.

Feature	This digital system	[BAG 09]	[VYK 09]
Technology	UWB-ISM (2.4 GHz)	UWB–UHF (900 MHz)	UHF (900 MHz)
Downlink data rate	40 kbps	40–160 kbps	6.8 kbps
Uplink data rate	Up to 10.1 Mbps	Up to 10 Mbps	
Logic clock	1 MHz	10 MHz	4 MHz
Supply voltage	3 V	2.75 V	3 V
DC current	Sleep: 20 nA Answer: 5/75 μA	Sleep: 1.5 μA Answer: 51 μA	Sleep: 0.6 μA Answer: 12 μA
Maximum distance	8.5 m	13.9 m	2,995 m (Measured 4.26 m)
Special features	Able to integrate I^2C sensors, ADC	UWB pulse generator	Paper substrate, EPC compliant

Table 4.2. *Comparison between digital semi-passive RFID systems*

5

Wireless Sensors Using Semi-passive UWB RFID

5.1. Introduction

This chapter presents the implementation of some wireless sensors with the semi-passive time-coded UWB RFID systems presented in Chapter 4. Four implementations are presented:

– a temperature sensor using the analog semi-passive UWB RFID system (see section 4.4), using a PIN diode at the end of the transmission line, and powered by solar energy;

– a nitrogen dioxide (NO_2) gas sensor using carbon nanotubes (CNTs). It is integrated on the analog semi-passive UWB RFID system (see section 4.4), with the backscatterer based on the RF switch;

– a multi-sensor tag using the microcontroller-based semi-passive UWB RFID system (see section 4.3), intended for smart city applications, and capable of measuring;

    - temperature,

    - humidity,

    - acceleration,

    - barometric pressure;

– a second nitrogen dioxide (NO_2) gas sensor, connected to the microcontroller-based semi-passive UWB RFID system (see section 4.3), which digitizes the CNT resistance.

This chapter is organized as follows:

– section 5.2 presents the solar-powered temperature sensor using the analog semi-passive UWB RFID system;

– section 5.3 presents the nitrogen dioxide gas sensor based on CNTs;

– section 5.4 presents the sensor integration in the microcontroller-based semi-passive UWB RFID system. The multi-sensor tag and the nitrogen dioxide sensors are presented;

– finally, section 5.5 compares chipless (see Chapter 3) and semi-passive time-coded UWB RFID sensors, and draws the conclusions.

5.2. Solar-powered temperature sensor based on analog semi-passive UWB RFID

5.2.1. *Introduction*

Several wireless temperature sensors have been recently presented. In [YIN 10], an EPC Gen2 tag with an embedded temperature sensor is proposed. In [KOC 06], another temperature sensor is integrated in a tag that uses a 2.3 GHz signal to send data and a 450 MHz signal to receive power. Wireless measurements of temperature are given in [VYK 09] for a sensing UHF RFID platform based on paper substrates. Finally, [CHO 05] presents another custom system with a temperature and photo sensor. In [YIN 10, KOC 06, VYK 09] and [CHO 05] cases, narrow band signals have been used. In addition, big efforts are being done during pastyears to power wireless sensors and RFID tags by means of green energies, such as solar energy [KIM 12, GEO 12].

Here, a temperature sensor based on time-coded UWB RFID is presented. The sensor is powered by solar energy. The PIN diode current is controlled by a negative temperature resistor (NTC). Then, the backscattered response of the tag is modulated in amplitude by the temperature.

5.2.2. Sensor design and calibration

Figure 5.1 shows a scheme of the sensor, based on the design from section 4.4. It consists of four main blocks. The PIN diode-based UWB backscatterer block (see section 4.4.3) enables the sensor to communicate with the reader. It is based on a broadband eccentric annular monopole antenna connected to a delay line. The delay line length L is chosen to separate the structural and tag modes with a round-trip delay of about 1 ns. Two slots between the antenna ground plane and the circuitry have been introduced, as explained in section 2.5.2. The detector and calibrator are explained in detail in section 4.4.4. Finally, there is a power supply circuit block that consists of a solar cell and a regulator that generates the bias voltages and voltage thresholds for the state comparators. Photographs of the fabricated sensor tag are shown in Figure 5.2. It is manufactured on Rogers RO4003C substrate (see Table 2.3). The tag size is 15.7 cm× 8.2 cm.

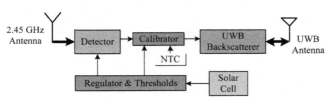

Figure 5.1. *Scheme of the solar-powered temperature sensor*

The calibrator circuit is based on the design described in section 4.4.4 (which is detailed in Figure 4.33(a)). Figure 5.3 shows in detail the part corresponding to the conditioning circuit, which delivers a 1OUT voltage to the COM1 input of the DC switch. The COM1 voltage depends on the value of the AVX NB20R00684 NTC SMD thermistor. This COM1 temperature-dependent voltage is generated from the non-inverting amplifier structure of the second TLV2402 amplifier B1. It has a gain of 2.2 set by the 220 and 100 kΩ resistors. The non-inverting input IN1+ is the voltage at the NTC in parallel with a 150 kΩ resistor, minus the voltage drop at the 820 kΩ series resistor. With these values, the output voltage 1OUT is between 0.6 V at 70°C (NTC = 420 kΩ) and 1.2 V at 35°C (NTC = 98 kΩ). It corresponds to the shaded zone of the PIN-diode backscatterer (see section 4.4.3).

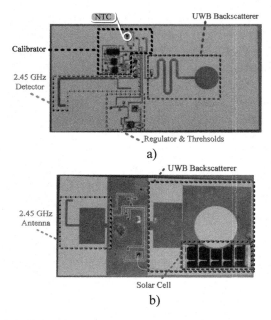

a)

b)

Figure 5.2. *Fabricated solar-powered temperature sensor, a) top face and b) bottom face. For a color version of the figure, see www.iste.co.uk/ramos/rfid.zip*

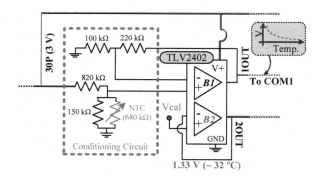

Figure 5.3. *Scheme of the signal conditioning circuit for the temperature sensor. For a color version of the figure, see www.iste.co.uk/ramos/rfid.zip*

5.2.3. Solar-cell integration: power requirements

The flexible amorphous silicon Power Film SP3-37 solar cell is selected. The specifications of SP3-37 are an open-circuit voltage $V_{oc} = 4.1$ V and a short circuit current $I_{sc} = 28$ mA, when illuminated by a light source complying with the standard AM1.5G global solar irradiance spectrum [AME 14], T = 25°C and 1 sun (= 100 mW/cm^2) irradiance.

The SP3-37, shown in Figure 5.4, is rectangular with dimensions 64 mm × 37 mm. In Figure 5.4,we can see the positive and negative terminals of the solar cell as well as a number of conductive strips on its top surface. The cell itself is a small solar module consisting of five smaller cells isolated by each of the four intermediate horizontal conductive strips that are electrically connected in series. Each of the individual cells provides an open-circuit voltage Voc = 0.82 V and short circuit current Isc = 28 mA, set by the width of the cell.

The integration of solar cells with antennas was originally proposed in [TAN 94]. The placement of a solar cell on top of a printed antenna does not affect the antenna performance provided that the area of the conductive surface of the antenna where the current density is high, such as near the feed point and the radiating edges, is not covered by the cell [TAN 94, COL 13]. This fact allows us to significantly reduce the total area required for the circuitry and solar cells, allowing for a more compact system implementation.

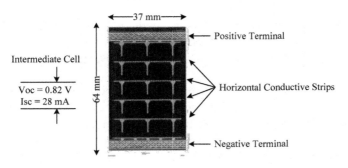

Figure 5.4. *Power film SP3-37 solar cell*

The solar cell used to power the sensor is therefore integrated on the UWB antenna ground plane. In order to avoid placing the solar cell near the antenna feed point and due to the presence of the circular disc aperture in the ground plane limiting the available area, it was necessary to cut the original cell along its length into two pieces capable to produce an open-circuit voltage V_{oc} = 4.1 V and a short circuit current I_{sc} = 14 mA under 1 sun irradiance. As is shown in Figure 5.5, it is necessary to further shape the two pieces in order to conform to the ground plane conductive area leading to a non-uniform width, which results in a slightly reduced current capability. The current capability of each solar cell piece alone is sufficient to power the sensor circuitry.

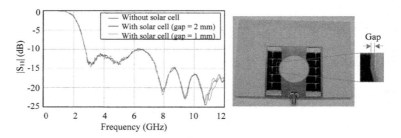

Figure 5.5. *Measured $|S_{11}|$ parameter of the tag antenna with and without the solar cell. For a color version of the figure, see www.iste.co.uk/ramos/rfid.zip*

Figure 5.5 shows the measured $|S_{11}|$ parameter of the UWB monopole with and without two solar cells. There are two measurements, one with the solar cells placed with a 1 mm gap between the circular slot and the cell and the other with a 2 mm gap. The antenna performance is not affected at all by the cells.

Each cell has sufficient current capability to power the sensor tag. As a result, only one cell is placed in the final circuit prototype as can be observed in Figure 5.2(b). It should be noted that the negative terminal of the cell is directly soldered on the ground plane conductor while the positive terminal of the cell is connected to the circuit supply using an insulated wire.

The output of the solar cell is connected to a Linear Technologies LT1763 (LT1763CS8#PBF) variable voltage regulator. As shown in Figure 5.6, it is adjusted to provide a 3 V output with two resistors of 500 and 750 kΩ, according to the manufacturer: $Vcc = 1.22(1 + R_2 / R_1) + I_{ADJ} R_2$, where $I_{ADJ} = 30$ nA typically. The thresholds needed for the detector and calibrator circuits are obtained using a simple resistive circuit.

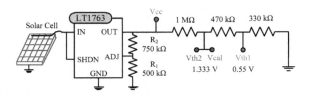

Figure 5.6. *Scheme of the solar cell connected to the regulator and the resistive values for the thresholds*

The tag needs to work by its own power source (in this case, solar energy). Since the consumption is a major concern in autonomous sensors, a study is carried out here. Table 5.1 shows the current consumption of each element in the tag circuitry. As expected, the most consuming element is the PIN diode, which draws up to 75 μA when it is in a calibration state. There are three possible combinations, depending on whether the tag is on a "Standby" state, waiting to receive the calibration signal by the reader, a "Temperature" ("Sensor" in section 4.4) state or a "Calibration" state. Their corresponding currents, I_{Stdby}, I_{Temp} and I_{Cal}, respectively, are calculated next. As shown, the maximum current consumption is around 82 μA.

$$I_{Stdby} = I_{Th} + 3I_{Cmp} = 4.30 \text{ μA} \qquad [5.1]$$

$$I_{Temp} = \begin{cases} I_{Th} + 5I_{Cmp} + I_{Sw} + I_{Dmm} = 22.06 \text{ μA (at 70 °C)} \\ I_{Th} + 5I_{Cmp} + I_{Sw} + I_{DmM} = 62.06 \text{ μA (at 35 °C)} \end{cases} \qquad [5.2]$$

$$I_{Cal} = I_{Th} + 5I_{Cmp} + I_{Sw} + I_{Dc} = 82.06 \text{ μA} \qquad [5.3]$$

Element	Symbol	Current (μA)	
		Minimum	Maximum
Threshold resistive divisor	I_{Th}		1.66
TLV240X	I_{Cmp}		0.88/channel
MAX4523	I_{Sw}		1
PIN Diode (Temperature)	I_{Dmm}, I_{DmM}	15 (at 70 °C)	55 (at 35 °C)
PIN Diode (Calibration)	I_{Dc}		75

Table 5.1. *Current consumption of the solar-powered temperature sensor by element*

It is important to note that the current consumption of this setup is noticeably lower than the consumption required by setups based on a microcontroller, such as the digital microcontroller-based approach presented in section 4.3. Even though that the tag in section 4.3 consumes 75 μA while answering, it requires 250 μA to operate the analog-to-digital converter (ADC) module. This module is required to acquire the output voltage from the conditioning circuit of the NTC. Therefore, the total consumption of the microcontroller-based approach is about 325 μA when using the ADC, which is higher than the 82 μA here. Other commercial microcontrollers such as the Texas Instruments MSP430 have lower ADC consumption (75 μA). However, considering the total consumption, it would sum up to around 150 μA, which is still above 82 μA.

5.2.4. *Results and error study*

To evaluate the validity of the system, the tag performance is measured using the hybrid reader described in section 4.1.2. Figure 5.7(a) shows the unprocessed (RAW) signals of the tag response at a 40 cm distance, heating the tag up to 70°C with a heat gun and letting it cool down on still air to 35°C. Figure 5.7(a) also shows the RAW signals for the background state (s_{back}), and the calibration state (where the PIN diode is biased with V_{cal}). A very large coupling contribution from the reader's Tx to Rx antenna is present before the tag response, since the separation between the reader's antennas is smaller than the reader–sensor distance. Also, some clutter is present due to reflections with surrounding objects. Figure 5.7(b) shows the zoomed area corresponding to the tag response. The structural modes, as shown, remain identical for all the possible temperatures of the tag. On the contrary, the tag modes' amplitudes change. The tag mode with the largest amplitude corresponds to the s_{back} state, which corresponds to the tag measurement without any current flowing through the diode, that is, without any 2.45 GHz calibration signal sent to the tag. With this background measurement, the clutter and coupling contributions can be diminished without having to measure the empty-room response.

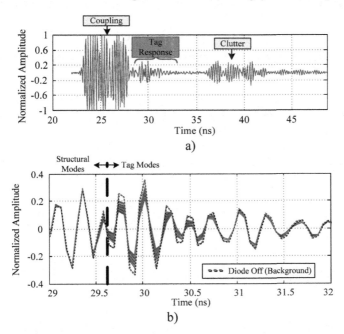

Figure 5.7. *a) Unprocessed time-domain signals for the tag response from 35 to 70°C, b) zoomed tag response*

To enhance the tag modes, for each measurement s_i (which contains the coupling contribution, the structural and tag modes, and clutter) the background s_{back} is subtracted as shown in Figure 5.8. It can be clearly observed that the coupling contribution and clutter have been diminished in front of the tag mode. Since the structural mode is also the same for the background (the tag is always present at the scene), it is also removed.

Figure 5.9 shows the background-subtracted signal from Figure 5.7 after applying the CWT. As shown in the inset of Figure 5.9, the amplitude grows when temperature changes from 35 to 70°C. All amplitudes are normalized with respect to the calibration state.

The calibration to temperature ratio can be defined as the ratio between the amplitude in the calibration state and the amplitude in the temperature measurement state. It can be obtained by detecting the peaks of the tag modes in Figure 5.9. Figure 5.10 shows the calibration to temperature ratio as a function of the diode current, which varies from 15 to 60 μA.

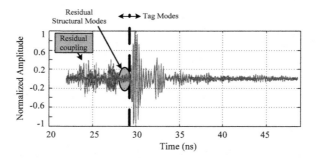

Figure 5.8. *Tag response after subtracting the background signal. For a color version of the figure, see www.iste.co.uk/ramos/rfid.zip*

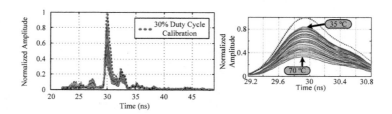

Figure 5.9. *Tag response after subtracting the background signal and applying the CWT. For a color version of the figure, see www.iste.co.uk/ramos/rfid.zip*

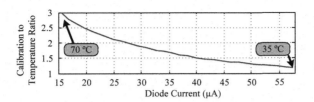

Figure 5.10. *Calibration to temperature ratio*
as a function of the diode current

The real temperature is obtained with a reference wired multimeter (Agilent 34410A) from the output voltage of the calibrator. Since the elements in the circuit are known, the equivalent NTC resistance R_{NTC} for a given voltage can be calculated. In this manner, the wireless performance is compared to the same NTC sensor when it is measured with a stable, wired instrument. The real temperature T for a given resistance can be obtained using the manufacturer's values (R_n = 680 kΩ at 25°C, β = 4,400, T_2 = 298 K):

$$T = \left[\frac{\ln\left(\frac{R_{NTC}}{R_n} \right)}{\beta} + \frac{1}{T_2} \right]^{-1} \quad [K]. \tag{5.4}$$

Figure 5.11 shows the calibration to temperature ratio for eight random measurements at distances tag-reader from 40 to 150 cm, as a function of the real temperature. The mean ratio for these eight measurements is also shown. As can be observed, all the measurements are very similar.

Figure 5.12 shows the real temperature as a function of the mean ratio from Figure 5.11, and a fifth-degree polynomial regression. The polynomial regression coefficients are used to obtain the estimated temperature from the ratios. This step is required only once in order to obtain the temperature calibration curve of the sensor.

Finally, Figure 5.13 shows the estimated (measured) temperature as a function of the real temperature for the eight measurements. The estimated temperature is calculated for each of the eight measurements using their ratios and the polynomial regression parameters from Figure 5.12. As observed, most of the measurements are very close to the real temperature, validating the functionality of the system, independently of the distance and angle.

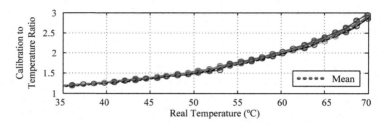

Figure 5.11. *Calibration to temperature ratio as a function of the real temperature for eight random measurements from 40 to 150 cm. For a color version of the figure, see www.iste.co.uk/ramos/rfid.zip*

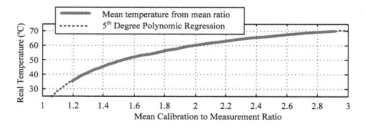

Figure 5.12. *Mean real temperature as a function of the calibration to temperature ratio and corresponding fifth-degree polynomic regression*

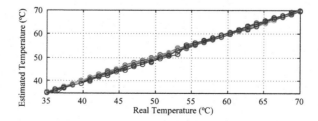

Figure 5.13. *Estimated temperature as a function of the real temperature for eight random measurements from 40 to 150 cm. For a color version of the figure, see www.iste.co.uk/ramos/rfid.zip*

For the purpose of characterizing the error obtained with the tag sensor, a study is carried out next. Figure 5.14(a) shows a histogram obtained from 500 measurements at a fixed temperature (29°C) and distance (40 cm). In this case, 88% of the measurements are under 0.6°C of error. Figure 5.14(b) shows another histogram for the eight random measurements of Figure 5.13. Tag-reader distances are between 40 and 150 cm and the temperature range is from 35 to 70°C. Here, 84% of the measurements are within a 0.6°C of error.

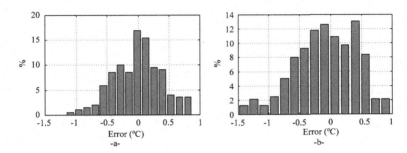

Figure 5.14. *Absolute temperature error as a function of the real temperature for a) 500 measurements at 29°C and 40 cm, and b) eight random measurements and temperatures (within 35–70°C) from 40 to 150 cm*

Finally, Figure 5.15 shows the mean relative error of Figure 5.14(b). In comparison with the chipless PTS temperature sensor (see section 3.2), these results are similar for the fixed temperature/distance cases. The main advantage of the sensor shown here is that the error for a fixed case is very similar to the error for the case of the sensor being moved. As explained in section 3.2, in chipless sensors the error is greatly increased when the distance or orientation between the tag and the reader changes, and they require a calibration curve for each tag–reader distance/orientation. This means that the approximate distance between the tag and the reader is needed: this is something impractical in real applications, and something that is not required in this sensor.

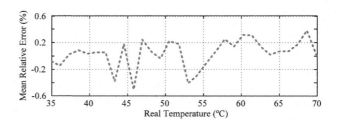

Figure 5.15. *Mean relative temperature error as a function of the real temperature for 8 random measurements and temperatures (within 35°C to 70°C) from 40 cm to 150 cm*

5.3. Nitrogen dioxide gas sensor based on analog semi-passive UWB RFID

5.3.1. *Introduction*

Wireless sensing of hazardous gases is a subject that has been under research recently [THA 11]. Some gases such as ammonia (NH_3), carbon dioxide (CO_2) and nitrogen dioxide (NO_2) are potentially hazardous for humans, as well as for livestock and agriculture. CNTs may be an enabling technology for detecting these gases since they provide a large specific surface area to interact with their environment, their electrical conduction changes dramatically upon gas absorption, and response and recovery of CNT sensors operated at room temperature has been reported [ZAN 11].

Several recent investigations have aimed to detect gases using CNT-based sensors integrated into passive or semi-passive RFID tags. Ammonia (NH_3) and nitrogen dioxide (NO_2) CNT-based sensors were presented in [LAK 11] for concentrations of 4 ppm NH_3 and 10 ppm NO_2 at both 864 MHz and 2.4 GHz measured on wired transmission lines and intended for use in RFID sensors. A multiwall CNT-based carbon dioxide (CO_2), oxygen (O_2) and NH_3 passive wireless sensor was presented in [ONG 02], based on an inductor-capacitor resonant circuit and providing measurements at a 15 cm reader–tag distance. A surface-modified multiwalled CNT-based sensor was presented in [ABR 04], detecting dichloromethane (CH_2Cl_2), acetone (C_3H_6O) and chloroform ($CHCl_3$), and can be measured remotely since it is connected to a Bluetooth module. Single-wall CNTs were used as an impedance loading on a conventional passive RFID 915 MHz tag in [OCC 11], detecting 6 mL of 10% of NH_3 at a reader–tag distance of 63.5 cm. A CNT-coated surface acoustic wave (SAW) CO_2 sensor has been designed to be integrated into a wireless sensor and was reported in [SIV 08]. An inkjet-printed CNT-based chipless RFID sensor was also proposed for CO_2 detection in [VEN 13], with measurements at 20 cm provided.

Here, a gas sensor based on time-coded UWB RFID is presented. The sensor uses CNTs as the transducer. The CNTs change its electrical resistance depending on the gas concentration, in this case, nitrogen dioxide (NO_2). The CNTs change the state of the RF backscatterer. The backscatterer used is the backscatterer-based on the RF switch from Skyworks, as explained in section 4.4.3.

5.3.2. CNT-based nitrogen dioxide sensor

The multiwall carbon nanotubes (MWCNT) were obtained from Nanocyl, S.A. (Belgium). They are synthesized by chemical vapor deposition and have a purity of over 95%. They measure up to 50 μm in length and their outer and inner diameters range from 3 to 15 nm and 3 to 7 nm, respectively. A uniform functionalization with oxygen is applied to the CNTs provided so as to improve their dispersion and surface reactivity. For this activation step, the MWCNTs are placed inside a glass vessel, and a magnet, externally controlled from the plasma chamber, is used to stir the nanotube powder during the plasma treatment. Inductively coupled plasma at a frequency of 13.56 MHz is used during the process. Once the MWCNT powder is placed inside the plasma glow discharge, the treatment is performed at a pressure of 0.1 Torr, using a power of 15 W, and the processing time is adjusted to 2 min. A controlled flow of oxygen is introduced into the chamber, which gives rise to functional oxygen species attached to the CNT sidewalls (i.e. oxygenated vacancies consisting of hydroxyl, carbonyl and carboxyl groups) [ION 06]. This functionalization has been found to enhance sensitivity toward NO_2 [ZAN 11].

In the second processing step, the functionalized CNTs are dispersed in an organic vehicle (dimethylformamide), ultrasonically stirred for 20 min at room temperature and then air-brushed onto the sensor substrate while the resistance of the resulting film during deposition is controlled. Controlling film resistance during deposition enables sensors with reproducible baseline values to be obtained [HAF 13]. Figure 5.16 shows an image of the CNTs after functionalization. The image is obtained with transmission electron microscopy (TEM). It shows that the morphology of the CNTs is preserved during the functionalization.

Figure 5.17(a) shows the manufactured CNT sensor. It is deposited on Rogers RO4003C substrate (see Table 2.3). It consists of an interdigital copper structure that series-connects input and output. The MWCNT is deposited over this interdigital structure. Figure 5.17(b) shows the gas chamber used to house the sensor, while Figure 5.17(c) shows a detail of the chamber cap with the flow tubes.

Figure 5.18 shows the measurement in DC (with an Agilent 34410A digital multimeter) of the CNTs for a 100 ppm of NO_2 flow. As can be observed, an absolute resistance variation of about 70 Ω is obtained.

Figure 5.16. *Typical TEM microscopy image recorded on CNTs after the oxygen plasma functionalization*

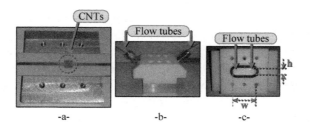

Figure 5.17. *Photographs of the substrate with the CNT sensor a), of the closed gas chamber with the access flow tubes b) and of the conduit with the open chamber c). Size of the chamber: w = 23.5 mm × h = 7.35 mm*

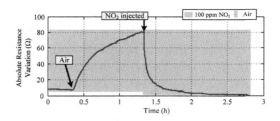

Figure 5.18. *DC measurement of the CNT absolute resistance variation for a 100 ppm of NO₂ flow. For a color version of the figure, see www.iste.co.uk/ramos/rfid.zip*

Next, the behavior of the CNT-based sensor is studied as a function of frequency. The S-parameters of the CNTs are measured in transmission (S_{21}) with an Agilent E8364C VNA. Then, by using the same calibration kit for the PTS chipless temperature sensor (see section 3.2.2.1), the S-parameters are de-embedded [AGI 04], and the resistance of the nanotubes is obtained.

Figure 5.19 shows the measurement of Figure 5.18 (repeated and connected to the VNA) as a function of frequency. The CNTs are measured for a 100 ppm of NO_2 flow for 300 kHz, 2.1748 MHz and 4.0496 MHz. The same shape as in the curve in Figure 5.18 curve is observed for 300 KHz; however, the resistance variation has been reduced to less than half of the DC case. If the frequency is increased up to several MHz, the variation decreases to less than 1–2 Ω.

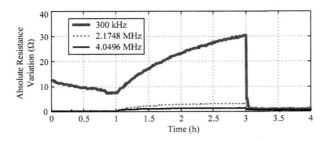

Figure 5.19. *CNT absolute resistance variation for a 100 ppm of NO₂ flow at 300 kHz, 2.1748 MHz and 4.0496 MHz. For a color version of the figure, see www.iste.co.uk/ramos/rfid.zip*

Figure 5.20 shows the same absolute resistance variation as a function of frequency and time. As it can be observed, as the frequency increases, the variation decreases even more. A range of about 50 Ω at the 1–3 GHz range is needed for chipless resistive sensors (see section 3.2). Given this, it is clearly noticeable that the sensor could not be read with a chipless tag as it is. Therefore, it is integrated into a semi-passive tag.

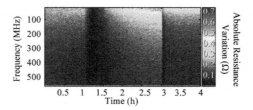

Figure 5.20. *CNT absolute resistance variation for a 100 ppm of NO$_2$ flow at frequencies between 0 and 550 MHz. For a color version of the figure, see www.iste.co.uk/ramos/rfid.zip*

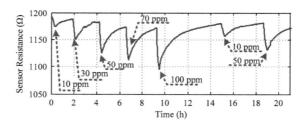

Figure 5.21. *Response and recovery cycles of a CNT sensor operated at room temperature. The evolution of sensor resistance for several concentrations of NO$_2$ is shown*

Finally, Figure 5.21 shows the evolution of sensor resistance (in DC) as a function of the NO$_2$ concentration. The first part consists of successive response-recovery cycles at increasing concentrations (10, 30, 50, 70 and 100 ppm) of NO$_2$. Each response-recovery cycle consists of 10 min of exposure to diluted NO$_2$ followed by 100-min recovery in dry air. A moderate baseline drift can be observed. However, the baseline is recovered in full when the duration of the cleaning cycle is increased to 140 min or more. Finally, the detection of 10 and 50 ppm of NO_2 is also shown. It is important to stress that response and recovery experiments were performed with the sensor always operating at room temperature.

5.3.3. *Wireless sensor design and calibration*

The sensor is based on the analog semi-passive UWB RFID with a switch-based UWB backscatterer (see section 4.4.2). The UWB backscatterer uses a Vivaldi antenna, as in section 2.5.2. Figure 5.22 shows a photograph of the tag. It is fabricated on a Rogers RO4003C substrate (see Table 2.3) and its size is 120 mm×120 mm. All the parts are labeled. The tag is powered by a 3 V lithium battery. Following the lines detailed in section 4.4.2, $L_2 = 0$ mm and the L_3 round-trip delay is 850 ps. This separation is enough to detect both tag modes, given the 100 ps time resolution demonstrated in section 2.6.1. For convenience, bearing in mind the measurements in the gas chamber, the CNT is placed on a separate board and wire-connected to the sensor tag, marked CNT in Figure 5.22(a). A reference connector (marked "Ref." in Figure 5.22(a)) is used to measure voltage V_1 using a multimeter for sensor verification and first-time calibration.

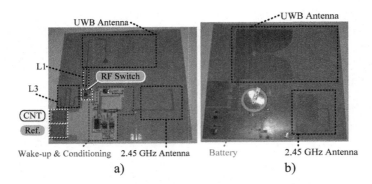

Figure 5.22. *Photograph of the fabricated tag, a) top and b) bottom. For a color version of the figure, see www.iste.co.uk/ramos/rfid.zip*

The calibrator circuit is based on the design described in section 4.4.4. Figure 5.23 shows in detail the part corresponding to the conditioning circuit, which delivers a 1OUT voltage to the COM1 input of the DC switch. The COM1 voltage depends on the value of the CNT resistive sensor. This COM1 gas-dependent voltage is generated from the non-inverting amplifier structure of the second TLV2402 amplifier B1. It is adjusted to provide an output between 1 and 1.4 V for the concentrations between 1,000 and 1,200 Ω of Figure 5.21. It corresponds to the shaded zone of the switch-based UWB backscatterer (see section 4.4.2).

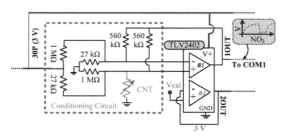

Figure 5.23. *Scheme of the signal conditioning circuit for the nitrogen dioxide sensor*

5.3.4. *Results*

Using the time-domain radar setup (see section 2.3.2), the sensor is measured at a sensor–reader distance of 1 m. The dimensions of the laboratory where the NO_2 line is available limit this distance. Figure 5.24 shows a photograph of the measurement room and the basement where the gas canisters are located.

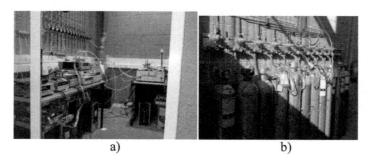

Figure 5.24. *a) Photograph of the gas measurement room. b) Photograph of the basement where the canisters are located*

To obtain the RF switch calibration curve, Figure 5.25 shows the unprocessed UWB signal as a function of V_1 for voltages from 0 to 3 V, driven with a wired programmable power supply and without the sensor connected. A large coupling contribution from the reader transmitter to the receiver antenna and clutter due to reflections from nearby objects in the scene can be observed, masking the tag response. In order to reduce these contributions, the time-domain signal measured with $V_1 = 0$ V is used as the background and subtracted from all measurements. Thus, the coupling and clutter contributions are subtracted because they do not depend on the tag mode. This enables a background subtraction to be performed before each measurement, and hence a time-variant background is not a problem for the sensor. The structural mode is also heavily reduced. This is where the importance of the second tag mode can be seen. The sensor must now be identified from the time difference between the tag 1 and tag 2 modes, since the structural mode is no longer useful for this purpose. Figure 5.26 shows the time-domain signal after the background subtraction and after the continuous wavelet transform has also been applied to reduce noise. All signals are normalized with respect to the case of $V_1 = 3$ V.

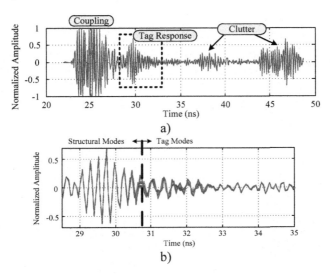

Figure 5.25. *Raw UWB signal as a function of V_1 a) and zoomed tag response b)*

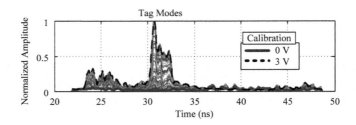

Figure 5.26. *Time-domain signals after background subtraction and after applying the CWT, normalized to the case of 3 V. For a color version of the figure, see www.iste.co.uk/ramos/rfid.zip*

Figure 5.27 shows the peaks of the tag 1 modes in Figure 5.26 as a function of V_1. The peaks are also normalized to $V_1 = 3$ V. An interpolation is calculated for the RF switch operation region using the piecewise cubic Hermite interpolating polynomial. This curve is used and stored as the sensor's calibration curve.

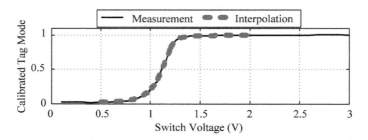

Figure 5.27. *Calibrated tag 1 modes as a function of V1. Interpolated calibration curve*

Now, the CNT sensor is connected to the tag. Three cycles consisting of 100 ppm of NO_2 for 10 min and air for 100 min are injected into the gas chamber. The results are shown in Figure 5.28. Using the calibration curve from Figure 5.27, the estimated voltage at the RF switch (V_1) is obtained from the calibrated tag mode in Figure 5.28(a) and shown in Figure 5.28(b). Voltage V_1 is also measured as a reference using an Agilent 34410A multimeter. This result is also shown overlapped in Figure 5.28(b). A repeatable pattern can be observed, and the wireless measurement is very close to the multimeter reference.

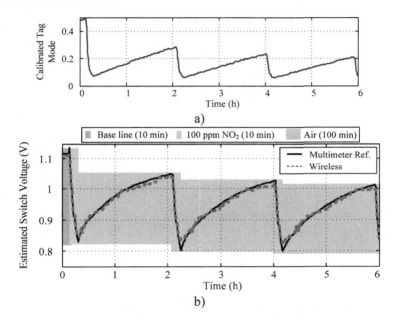

Figure 5.28. *a) Calibrated tag mode as a function of time for a varying CNT sensor; b) estimated switch voltage V_1 as a function of time for three measurement cycles of 100 ppm NO_2 compared to measured voltage V_1. For a color version of the figure, see www.iste.co.uk/ramos/rfid.zip*

In order to detect several concentrations of NO_2, the measurement profile from section 5.3.2 (see Figure 5.21) is chosen. The results are shown in Figure 5.29. Figure 5.30 shows the relative error of the wireless measurement with respect to the multimeter reference as a function of time. The mean relative error is 0.34%. To derive the sensitivity of the sensor, Figure 5.31 shows the relative resistance change obtained from the measurements in Figure 5.29. Relative changes of 3.9, 9.2, 14, 16.6 and 19.9% are obtained for 10, 30, 50, 70 and 100 ppm, respectively.

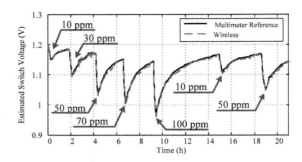

Figure 5.29. *Estimated switch voltage as a function of time for 10, 30, 50, 70 and 100 ppm concentrations of NO$_2$ compared to the wired reference*

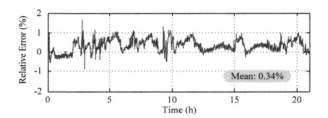

Figure 5.30. *Relative error between the wired reference and the wireless measurement*

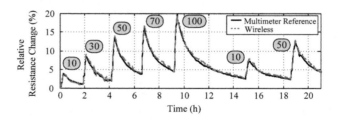

Figure 5.31. *Relative resistance change as a function of NO$_2$ concentration*

5.4. Sensor integration in microcontroller-based semi-passive UWB RFID

An alternative for wireless UWB sensors with the digital microcontroller-based approach is shown here. As explained in section 4.3.6, the digital tag can integrate both analog and digital sensors easily. To this end, two systems are shown:

– a multi-sensor tag that integrates analog DC and digital I2C sensors, for smart cities applications;

– a nitrogen dioxide (NO_2) digital sensor tag.

5.4.1. *Multi-sensor tag*

The block diagram of the multi-sensor tag is shown in Figure 5.32. It consists of a core logic based on the PIC 16F1827, a wake-up circuit to detect the signal coming from the 2.45 GHz modulator, a UWB backscatterer, which performs the communication between the tag and the reader and the sensor module. This module can be divided into two parts, those sensors that are connected directly to the ADCs of the PIC and those sensors that are connected to the PIC by means of an I2C interface. The sensors connected to the microcontroller ADC consist of a Texas Instruments TMP20 temperature sensor and a HoneyWell HIH-5031 humidity sensor. The sensors connected to the I2C bus consist of a FreeScale MMA8453QT accelerometer and an MPL115A2T1 barometer. A Microchip 24LC256 non-volatile EEPROM memory is also connected to the I2C bus to store measurements. This is a key feature in those applications where the sensor and the reader are not in continuous connection. It permits us to store measurements that are downloaded when the sensor and the reader are under connection.

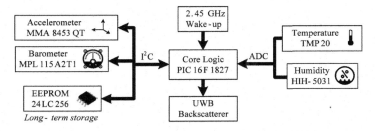

Figure 5.32. *Block diagram of the multi-sensor tag*

Figure 5.33 shows a vertical cut and photographs of the sensor tag. Its size is 12 cm × 6 cm and it is fabricated on Rogers RO4003C (see Table 2.3). It is powered using a 3 V Lithium battery. As shown in the vertical cut, the multi-sensor module is manufactured on a separate board that is overlapped to the tag board, as also shown in Figure 5.33.

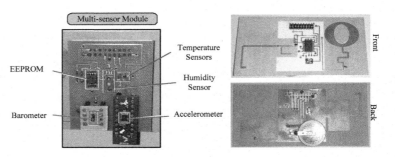

Figure 5.33. *Vertical cut of the multi-sensor tag and photographs of the front and back faces of the tag, and of the multi-sensor module board*

As an example, the measurement of temperature with the TMP20 sensor and of acceleration with the MMA8453QT sensor is shown next. Figure 5.34 shows a measurement of the differential signal (see section 4.3.4) as a function of time. The inset shows the signal prior to applying processing. On top of this, the CWT (see section 2.4.2) is applied to increase the signal-to-noise ratio.

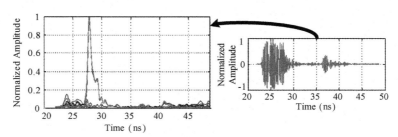

Figure 5.34. *Differential signal of a multi-sensor tag read after applying processing. In the inset, RAW signal before processing. For a color version of the figure, see www.iste.co.uk/ramos/rfid.zip*

Figure 5.35 shows the signal sequences corresponding to two measurements of temperature at 50 cm. The first measurement (top) is 24.91°C, and the second is 108.79°C. Temperature has been also obtained by wired means as a reference to validate the system. The measured values are 24.516 and 107.8474°C. The reference of the MMA8453QT accelerometer is read using the FreeScale Sensor Toolbox. The orientation bits from the PL_STATUS register are considered: LAPO and BAFRO. LAPO indicates whether the sensor is in a landscape or portrait orientation. BAFRO indicates if the sensor is in a back or front orientation. The accelerometer functionality is tested by just changing the sensor position. Figure 5.36 shows a photograph of the sensor next to the read for the front, portrait up (LAPO= "00", BAFRO= "0") orientation.

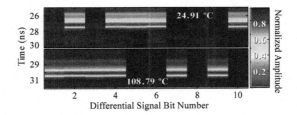

Figure 5.35. *Differential signal sequence for two temperature measurements. For a color version of the figure, see www.iste.co.uk/ramos/rfid.zip*

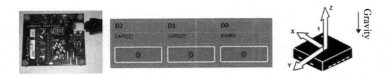

Figure 5.36. *Accelerometer read for the front, portrait up orientation*

5.4.2. *Nitrogen dioxide gas sensor*

Based on the same CNT gas resistive sensor presented in section 5.3, an alternative of the same sensor integrated in a digital tag is shown next. Figure 5.37 shows a photograph of the manufactured tag. It uses the same backscatterer and monopole configuration of section 4.3.

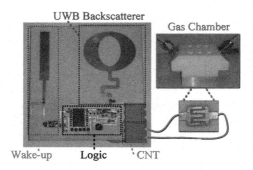

Figure 5.37. *Photograph of the manufactured digital nitrogen dioxide sensor tag*

Figure 5.38 shows the circuit scheme of the tag with the conditioning circuit to convert the resistance to a voltage drop. The same conditioning circuit presented in section 5.3.3 is used (with a single TLV2401 operational amplifier). In this case, the conditioning circuit output is connected to the PIC's ADC(RA1). The range of operation of this circuit (between 1 and 1.4 V) falls inside the ADC's requirements (0 to 3 V). Moreover, if the sensor resistance baseline drifts, which may happen during a long-time measurement (see section 5.3.4), the output values still have room to reach one of the boundaries (0 or 3 V). The conditioning circuit is powered from the PIC itself, using a digital output (RA4). This is an advantage, since the sensor only drains power when it is being read by the tag.

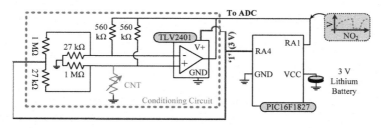

Figure 5.38. *Scheme of the circuit between the PIC16F1827 and the CNT sensor*

The sensor is operated at room temperature (i.e. 25°C) throughout the whole test procedure. The moisture level is kept constant at 10% relative humidity (R.H.) The test exercise involved four CNT sensors (to assess sensor reproducibility) and a measurement period of over 250 h at the end of which, sensors remained fully functional. The baseline resistance of the sensors (@25°C and 10% R.H. in air) was about 1,400 Ω. Figure 5.39 shows a comparison between the response of a wired measurement (with an Agilent 34410A multimeter in parallel) and a wireless measurement. The wired measurement is obtained only once and is used for comparing the performance of the wireless system. The actual physical parameter that the sensor changes is its electrical resistance (R_{CNT} in Ω). The following conversion factor can be applied to the detected voltage V_{ADC} in order to obtain the resistance: $R_{CNT} = 414.07 \times V_{ADC} + 696.88\ \Omega$. The margin 1.25 – 1.7 V from Figure 5.39 corresponds to, approximately, 1215 – 1400 Ω. The concentrations of NO_2 applied to the sensor in each measurement are marked out.

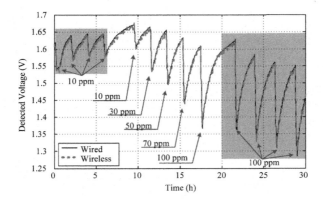

Figure 5.39. *Wired (black solid line) and wireless (red dashed line) measurements of the CNT-based NO₂ sensor with microcontroller. From left to right, four response-recovery cycles at 10 ppm, cycles at 10, 30, 50, 70 and 100 ppm, and four cycles at 100 ppm. The sensor is operated at room temperature (i.e. 25°C) both in the response and the recovery phases. For a color version of the figure, see www.iste.co.uk/ramos/rfid.zip*

Figure 5.40 shows the absolute error between the wired reference and the wireless measurement. The mean error is also shown, with a value of 4.5 mV. Theoretically, the ADC of the tag (a 10-bit resolution ADC embedded in a Microchip PIC16F1827 microcontroller) has a quantization error *QErr.* of:

$$QErr. = \frac{(V_{CC} - GND)}{2^{10}} = \frac{(3-0)}{1024} = 2.93 \text{ mV} .$$ [5.5]

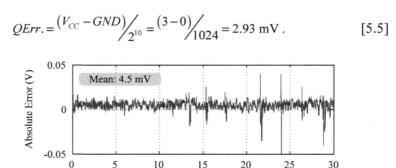

Figure 5.40. *Evolution of the absolute error between the wired measurement (taken as reference) and the wireless measurement*

The mean error is therefore close to the theoretical limit. The relative error remains mostly below 1%, never exceeds 2% and its mean value is 0.29%. Although some baseline drift can be observed in the response of the sensor to high nitrogen dioxide concentrations (e.g. see the repeated response-recovery cycles to NO_2 100 ppm in Figure 5.39), the baseline is regained without heating the sensor (i.e. at room temperature) if the sensor is given more time for recovery. Given the slow recovery of the sensor, which is due to the strong binding of NO_2 to the surface of CNTs, in a real application the detection phase would be run at room temperature and the use of a short temperature heating pulse would be needed at the beginning of the recovery phase in order to promote a faster return to baseline [LEG 10]. This would also suppress drift.

5.5. Comparison between chipless and semi-passive approaches: conclusions

Next, a comparison between chipless and semi-passive approaches is discussed. The chipless temperature sensor (section 3.2.2), the semi-passive digital temperature sensor (section 5.4.1), the semi-passive analog temperature sensor (section 5.2) and the semi-passive analog gas sensor (section 5.3) are compared in terms of cost and performance. Some elements are common in all approaches: a UWB radar used as reader and Rogers RO4003 substrate used to manufacture the tags. Therefore, the error of the UWB radar has been considered in the chipless approach (see Figure 3.8), and the costs of the reader and the substrate are not considered. All approaches have been summarized in Table 5.2.

Feature	Approach			
	Passive (chipless)	Semi-passive (digital, microcontroller)	Semi-passive (analog, PIN diode)	Semi-passive (analog, RF switch)
Complexity	Low	High	Medium	
Power supply	-	3 V battery	3 V battery or solar cell	
Power consumption	0 µA	30 nA (sleep) 400 µA (active, with ADC)	4.3 µA (sleep) 150 µA (active)	4.3 µA (sleep) 105.5 µA (active)
Suitable for hazardous environments	Yes	No	No	
Cost	Very low	Medium–high	Medium	
Error (°C)	3.5°C	0.14°C	0.6°C	-
ID Space	~8 bit	Configurable	~8 bit	
Read range	<2 m	8.5 m	8.5 m	
Calibration	Very complex	Simple	Complex	

Table 5.2. *Comparison of chipless and semi-passive approaches*

The main advantage of the chipless approach is its simplicity. The tag does not require its own power supply (battery), and can be used in hazardous environments such as chemical processes or vacuum chambers. The cost of the tag is small, and it is mainly attributed to the PTS sensor.

However, the error in the chipless tag is significantly higher than the error in the semi-passive tags. The chipless tag also requires a pre-stored calibration curve for each tag-reader distance and angle in order to minimize the error (see section 3.2.2). In addition, a background measurement must be performed before placing the tag, which could be impractical in some cases. The number of bits to code the ID of the chipless tag is small (limited by the time resolution of the system (see section 2.6.1). Finally, the read range of a chipless tag is below 2 m, even less when it is desired to accurately measure the tag mode amplitude variation.

The semi-passive digital approach permits us to code a large number of IDs, since the tag responds a binary sequence that can be programed. Moreover, since the tag has logic (microcontroller), more complex protocols can be implemented. For instance, data encryption algorithms can be implemented for secure data transmission. The error in temperature is very small. The semi-passive tag does not require either a complex calibration curve or a background measurement. It is, therefore, more reliable and practical than the chipless tag. The read range of the semi-passive tag obtained with this system is longer than 8 m.

The cost of the semi-passive digital approach is much higher than the chipless or even semi-passive analog approach. It requires a microcontroller, which limits its minimum price. For semi-passive approaches, the reader needs a 2.45 GHz oscillator, a power amplifier and a microcontroller in order to send the wake-up commands. The battery at the tag limits it to be used in hazardous environments. Moreover, there is power consumption due to the elements at the tag. In sleep mode, the only consuming element is the microcontroller, which consumes 30 nA (see section 4.3). In active mode, however, the microcontroller consumes 75 μA, added to the PIN diode when it is "on", which consumes 70 μA, and the TMP20, which accounts for 4 μA, and about 250 μA of the ADC, totaling a 400 μA consumption of the tag. This value, even though it is considerably low for a semi-passive tag given its features, contrasts with the zero value of the chipless approach. The semi-passive analog approaches require less power because no ADC is used. However, the standby current is larger.

In conclusion, depending on the application and its constraints (precision, number of sensors deployed, hazardous environment, distance), it will be more suitable to use one approach or the other.

6

Active Time-coded UWB RFID

6.1. Introduction

The read range of about 8 m achieved with the backscattered semi-passive systems in Chapter 4 might not be enough for long-range RFID-enabled localization applications. It is clearly noticeable that greater distances are required to cover a large indoor/outdoor area.

This chapter presents two active time-coded UWB RFID systems for localization applications. The goal is to increase the radar cross-section (RCS) of the tag by providing it with gain. This has already been used in active reflectors proposed as active radar calibrators [BRU 84]. In the same direction, passive and active Van Atta arrays have also been proposed for narrowband RFID applications [VIT 10, CHA 11].

Here, both systems consist of an amplifier integrated in a time-coded UWB RFID tag, which increases the power of the backscattered tag mode:

– a system based on an amplifier and a UWB RFID tag in cross-polarization;

– a system based on a reflection amplifier.

They will be presented and compared in terms of tag size, read range, cost and power consumption. This chapter is organized as follows:

– section 6.2 presents the active system based on an amplifier in cross-polarization;

– section 6.3 presents the active system based on a reflection amplifier;

– finally, section 6.4 compares both systems in terms of power consumption, bandwidth, read range and tag size.

6.2. Active UWB RFID system based on cross-polarization amplifier

Here, the tag consists of one receiver UWB antenna followed by a UWB amplifier connected through a delay line to a transmitter UWB antenna. This tag is simpler than other active tags or reflectors [WEH 10, DAR 04, RAB 04, BAG 09], because it does not require complex synchronization techniques between the pulse generator included in the tag (there is no pulse generator, only the reader's pulse is amplified) and the receiver in the reader [BAG 09]. Several low-noise integrated amplifiers for UWB have been reported in the literature (a comparison is given in [ZOU 11], Table III) and they can be integrated in the tag. These amplifiers achieve power consumption between 1.3 and 25 mW, depending on the Complimentary Metal-Oxide Semiconductor (CMOS) process, topology and gain. The tag shown here, therefore, needs no special integrated designs. In this section, a proof of concept prototype using commercial components is presented. In addition, since the read range is a key point in active tags, the link budget is also studied.

6.2.1. Introduction

Figure 6.1 shows a scheme of the system based on a cross-polarization amplifier. It comprises the reader and the tags. The reader interrogates the tag using a UHF (868 MHz) link, which wakes up the tag logic circuitry (a PIC 16F1827). The reader, logic circuitry, detector, wake-up system and protocol in the tag are based on the system discussed in sections 4.1–4.3. In the reader UHF end, a variable attenuator (NVA2500V35 from minicircuits) is used to control the transmitted power and to modulate the signal. Finally, the signal is amplified using a buffer amplifier (Gali84+ from minicircuits), followed by a power amplifier (RF3809 from RF monolithic solutions).

After the tag is woken, the reader sends a linearly polarized pulse to the tag, using the time domain radar from section 2.3.2. The pulse is received by one of the tag's UWB antennas. Then, the pulse is amplified and backscattered to the reader in the orthogonal polarization.

The backscattered signal is amplified or not in order to code a logical "0" or "1", as performed in section 4.3.3 by biasing the PIN diode. It is important to note that in this system the wake-up link frequency has been changed from the 2.4 to 2.5 GHz ISM band (section 4.2) to the UHF band, with the aim of increasing the wake-up distance. This is simply achieved by moving to a lower frequency (less attenuation in free space propagation), and working in a band where regulations permit the transmission of more power.

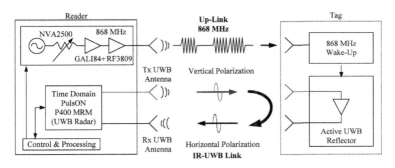

Figure 6.1. *Scheme of the UHF-UWB reader-tag system based on cross-polarization amplifier*

6.2.2. Cross-polarization amplifier design

Figure 6.2 shows the block diagram of the tag and Figure 6.3 shows a photograph of the tag and the experimental setup. The tag size is 133 mm by 133 mm. It is fabricated on Rogers RO4003C substrate (see Table 2.3). A commercial MMIC from Avago (ABA31563) is used here as amplifier. The amplifier consumes

13 mA at 3 V (39 mW). A MAX4715 switch from Maxim is connected to the amplifier DC supply. The switch is used as a buffer because the PIC's digital ports cannot support the required DC current by the amplifier.

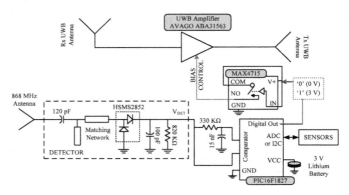

Figure 6.2. *Block diagram of the tag based on an active reflector*

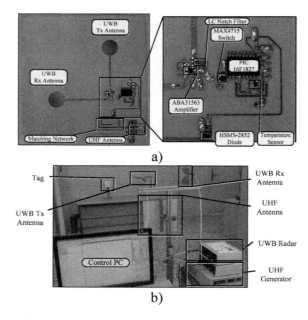

Figure 6.3. *Photograph of a) the implemented tag and b) the experimental setup*

Figure 6.4 shows the communication protocol used. It is similar to the protocol presented in section 4.3.5; first, the reader wakes the tag up using the UHF link. Then, the reader sends a UWB pulse and receives the backscattered answer. Next, it sends a new command to the tag via the UHF link to request the following bit, and so on. Note that the time-of-flight (TOF) is measured by the radar. No special synchronization between the tag and the reader is therefore required. This point is an important difference compared to other Impulse-Radio (IR)-UWB tags that use a pulse transmitter implemented in the tag instead of an amplifier [DAR 04, KRI 10]. The pulse transmitter must be synchronized with the reader in these systems. In addition, this tag is compatible with any UWB receiver (carrier based or IR based) because it acts as a wideband active reflector. The DC power consumption of the tag is comparable to or lower than other wireless systems such as Zigbee, but the advantage of this system is that it can also be combined with localization techniques based on TOF.

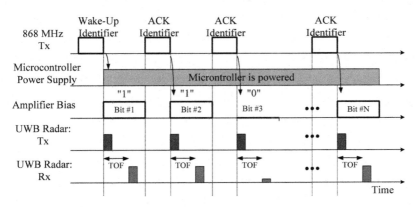

Figure 6.4. *Communication protocol for the active UHF-UWB RFID system*

Two broadband eccentric annular monopole UWB antennas (see section 2.5.1), placed in orthogonal positions, are used in the tag. The outer and the inner diameters are 36 and 18 mm respectively. The ground plane dimensions are 70 mm width per 50 mm height. The simulated reflection coefficient (S_{11}), gain and coupling (S_{21}) are shown in Figure 6.5. The separation between the antennas is established at 40 mm from the feed points in order to reduce coupling between them below −20 dB in the amplifier band. The simulations were performed with Agilent Momentum. The coupling between the cross-polarized antennas is low in the tag and the reader, and the clutter level reduction is important. However, the use of cross-polarized antennas has the drawback that the tag must be well oriented to the reader to prevent loss of signal by depolarization. In the applications where the tag orientation is unknown, circularly polarized antennas should be used in the tag or the reader end. Standard circularly polarized patch antennas such as the antennas used in commercial UHF RFID applications [CHE 09] can be used for the UHF link. Circularly polarized UWB antennas such as the antennas presented in section 2.5.3 could be used in the UWB link.

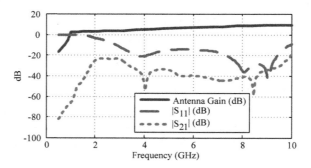

Figure 6.5. *Simulated antenna gain, reflect coefficient and coupling between Tx and Rx tag antennas*

In order to save space, a ceramic antenna (model P/N 0868AT43A0020 from Johanson Technology) was connected to the input of the UHF detector. The antenna has a peak gain of −1 dB in the 868 MHz band. A matching network composed of a 15 nH series inductor and 4.7 pF parallel capacitor was designed to match the antenna in the 865–868 MHz band.

Figure 6.6 shows the measured S parameters of the MMIC ABA31563 for the bias point of 3 V and 13 m A. Figure 6.7 shows the open loop gain computed from the cascade simulation of the amplifier and the antenna coupling. It can be seen that the open loop gain is up to 20 dB larger around 2 GHz due to the high gain of the amplifier. If necessary, the separation between antennas could be increased to minimize coupling and therefore to avoid oscillation, but the tag size would also increase. Another solution, shown here, is to insert a notch filter to reduce the gain. A simple LC series notch filter is implemented with a transmission line (4 mm long and 0.5 mm wide) and a 3 pF capacitor. The notch filter can be easily tuned to 2 GHz in order to reduce the gain, as shown in Figure 6.7.

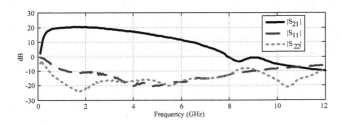

Figure 6.6. *Measured S parameters of the amplifier*

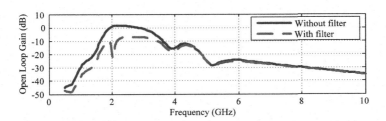

Figure 6.7. *Open loop gain without filter (solid line) and with notch filter (dashed line)*

The amplitude of the modulated reflected pulse is a function of the RCS of the reflector. The RCS of the active reflector can be computed using [BRU 84]:

$$\sigma = \frac{\lambda^2}{4\pi} G_{\text{tag},r} G_a G_{\text{tag},t}, \qquad [6.1]$$

where λ is the wavelength, $G_{\text{tag},r}$ and $G_{\text{tag},t}$ are the tag receiver and transmitted antenna gains respectively, and G_a is the amplifier gain including the notch filter.

Figure 6.8 shows the RCS computed using equation [6.1] with and without the filter. As expected, a reduction of the RCS around the center frequency of the notch filter is achieved, which decreases when the frequency moves away from the filter center frequency. For instance, a loss due to the filter of 1.4 dB at 4 GHz (which is around the 4.3 GHz center frequency of the reader's UWB pulse) is shown. At 4 GHz, an RCS peak of −7 dBsm is obtained and the RCS is almost flat around the center frequency of the pulse spectrum. The increase in the RCS compared to a passive reflector or tag is clearly visible.

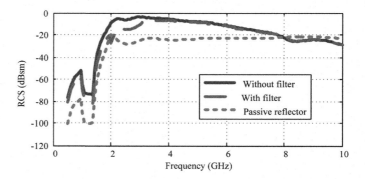

Figure 6.8. *Simulated RCS as a function of the frequency without (solid line) and with the notch filter (dashed line)*

6.2.3. *UWB and UHF link budget*

Concerning the UWB link, the bit error rate (BER) of the UWB link depends on the ratio between the energy per bit to noise spectral density (E_b/N_0). UWB radars can improve the E_b/N_0 by using coherent integration of pulses. If N_s pulses are integrated every T_b seconds, the bit rate can be written as rate = PRF/N_s, where PRF is the pulse repetition frequency. The signal-to-noise ratio (*S/N*) can be expressed as [DIB 06]:

$$\frac{S}{N} = \frac{\dfrac{E_b}{T_b}}{N_0 B} = \left(\frac{E_b}{N_0}\right)\frac{PRF}{N_s B},$$

[6.2]

where B is the bandwidth of the system (defined by the bandwidth of the UWB pulse and by the frequency response of all of the antennas). Then, the E_b/N_0 in dB can be expressed as:

$$\left(\frac{E_b}{N_0}\right)_{dB} = \left(\frac{S}{N}\right)_{dB} + PG(dB), \qquad [6.3]$$

where the processing gain (PG) is defined as:

$$PG\ (dB) = 10\log\left(\frac{B}{PRF}\right) + 10\log(N_s). \qquad [6.4]$$

The PG depends on the bit rate and the pulse repetition rate, which are parameters that depend on the target application. The noise spectral density N_0 is a function of the noise factor; according to the specification from the radar manufacturer, it is 4.8 dB [TIM 14]. An $F = 5.8$ dB has been chosen, adding 1 dB to take into account the cables between the radar and the antennas.

$$N_0 = kT_0F, \qquad [6.5]$$

and the average receiver power can be obtained from the radar equation:

$$S = \frac{P_t}{4\pi r^2}G_t\sigma\frac{1}{4\pi r^2}\frac{\lambda^2}{4\pi}G_r, \qquad [6.6]$$

where the RCS (σ) of the active reflector is given by [6.1], r is the tag-to-reader distance, P_t is the transmitted power, G_t and G_r are the gains of the Tx and Rx antennas and λ is the wavelength. Figure 6.9 shows that E_b/N_0 is calculated with the equations above. For the simulations, an amplifier with a gain of 16 dB, bandwidth $B = 1.35$ GHz [TIM 14], two UWB tag antennas with a gain of 5 dB and two UWB reader antennas with a gain of 6 dB were considered. The calculation was performed at the radar center frequency (4.3 GHz, see section 2.3.2) of the pulse and a 1.35 GHz bandwidth was taken into account. The average equivalent radiated power used is − 14.5 dBm (FCC 15b compliant) and the PRF is 10.1 MHz. Assuming a

BER limit of 10^{-3}, an E_b/N_0 of 6.8 dB is required for binary phase shift keying demodulation. From this simulation, depending on the number of samples N_s (1, 64, 4,096, or 32,768 can be used with the P400 radar), read ranges of 3.6, 9.5, 27.5, and 45.5 m can be theoretically achieved, respectively. To display the worst case, Figure 6.10 shows the E_b/N_0 calculated assuming the tag antenna gain is equal to 0 dBi. A comparison with the case of a passive tag ($G_a = 1$) [D'ER 12] is also included. In this case, the active approach indicates a read range of about 25 m compared with 12 m of the passive approach.

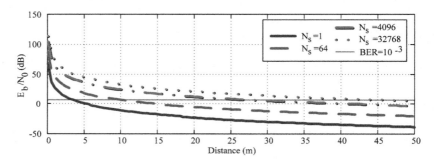

Figure 6.9. *Simulated E_b/N_0 as a function of the distance for different integration sampling numbers (Ns = 1, 64, 4,096, and 32,768). The limit line for BER = 10^{-3} (E_b/N_0 = 6.8 dB) is also indicated*

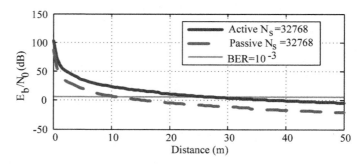

Figure 6.10. *Simulated E_b/N_0 as a function of the distance for an active reflector (solid line) and a passive reflector (dashed line). The tag antenna gain is assumed to be 0 dBi and the integration sampling number Ns = 32,768. The limit line for BER = 10^{-3} (E_b/N_0 = 6.8 dB) is also indicated*

Concerning the UHF link, Figure 6.11 shows the measured voltage at the output of the detector as a function of the input power. The tangential sensitivity (TSS) is the lowest input power for which the detector will have an 8 dB signal-to-noise ratio at the output of a test video amplifier. A TSS value of −57 dBm is typically obtained using zero bias diodes [AVA 14]. Setting the sensitivity of the wake-up system to −40 dBm (3 mV of detected voltage) would therefore be enough to ensure a good link quality. Figure 6.12 shows the received input power at the detector as a function of the distance. To avoid the uplink limiting the read range, the minimum wake-up distance should be larger than 45.5 m in free space, which is the theoretical simulated limit for the UWB link obtained above. As explained in section 4.2.2, a detector at the 2.4–2.5 GHz ISM band cannot achieve this distance considering power regulations. The UHF band was chosen for this reason. Using the maximum power transmitted under European (ETSI EN 302 208) and USA (FCC-15) RFID regulations, a theoretical distance longer than 70 m can be obtained in free space. A wake-up antenna gain of −1 dB was considered in the calculations. Since the wake-up distances are much larger than the maximum read range (limited by the UWB link), there is consequently a fading margin to mitigate multipath effects in the UHF channel. The data rate is limited by the low-pass filter at the output of the detector and by the speed of the internal comparator in the microcontroller. In the experiments, the data rate used is 1 kHz. However, it can be increased up to 40 kHz (see section 4.3.2). These speeds are high enough to send small configuration commands if sensors are integrated in this tag.

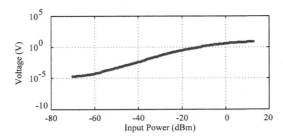

Figure 6.11. *Measured detected voltage as a function of the input power at 868 MHz*

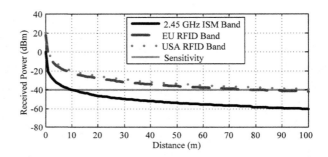

Figure 6.12. *Received power as a function of the tag-to-reader distance for the uplink*

6.2.4. *Results*

Figures 6.13–6.14 show the UWB signal received after background subtraction is applied as a function of the delay for a tag located 1 m and 10 m away from the reader, respectively, in an indoor scenario. A small reflection due to the tag structure is observed when the amplifier is biased OFF (OFF state). In contrast, there is a strong reflection when the amplifier is biased ON (ON state). A small delay is observed between the two states due to the propagation delay through the amplifier. Multiple reflections are also observed for the ON state due to feedback coupling between the tag antennas.

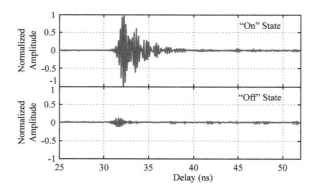

Figure 6.13. *Normalized received UWB raw signal as a function of time delay for the two states (amplifier ON and OFF) at a tag-to-reader distance of 1 m*

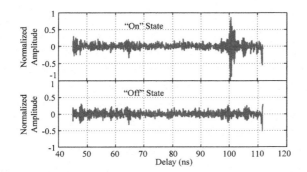

Figure 6.14. *Normalized received UWB raw signal as a function of time delay for the two states (amplifier ON and OFF) at a tag-to-reader distance of 10 m*

Figures 6.15–6.16 show the normalized magnitude of the continuous wavelet transform (CWT) (see section 2.4.2) applied to the received signal at 1 and 10 m, respectively. In both cases, the received signal in the ON state is clearly detectable for both distances. For the OFF state , the tag is not retransmitting the reader's pulse. The signal received therefore consists of background and residual clutter.

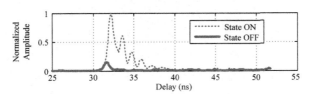

Figure 6.15. *Normalized maximum of CWT as a function of time delay for the two states (amplifier ON and OFF) at a tag-to-reader distance of 1 m*

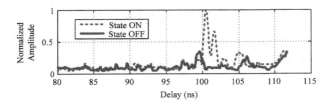

Figure 6.16. *Normalized maximum of CWT as a function of time delay for the two states (amplifier ON and OFF) at a tag-to-reader distance of 10 m*

Next, the differential coding scheme from section 4.3.4 is applied. Figures 6.17–6.19 show the maximum magnitude of the CWT applied to the received signal after subtracting the received signal for the first bit. An arbitrary bit sequence (0101011) is transmitted, which could correspond to a sensor data connected to the microcontroller, or an identifier code. In order to verify the simulations discussed in section 6.2.3, a measurement in an outdoor environment has been performed at a 45 m tag-reader distance. The result for the same sequence is shown in Figure 6.19. The effect of noise and clutter is clearly visible; however, the bit sequence can be demodulated. The time delay of the peak for the ON states depends on the time-of-flight delay between the tag and the reader, added to a small systematic delay between the reader and its own antennas. This small delay due to the reader's antennas is known and is always the same for any tag-to-reader distance. This active reflector could thus be combined with time-of-arrival localization techniques by using multiple readers located at different known positions (anchors).

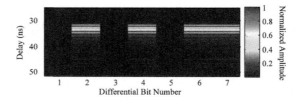

Figure 6.17. *Normalized maximum of CWT as a function of time delay for a bit sequence at a tag-to-reader distance of 1 m (indoor environment). For a color version of the figure, see www.iste.co.uk/ramos/rfid.zip*

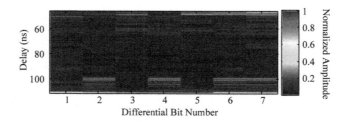

Figure 6.18. *Normalized maximum of CWT as a function of time delay for a bit sequence at a tag-to-reader distance of 10 m (indoor environment). For a color version of the figure, see www.iste.co.uk/ramos/rfid.zip*

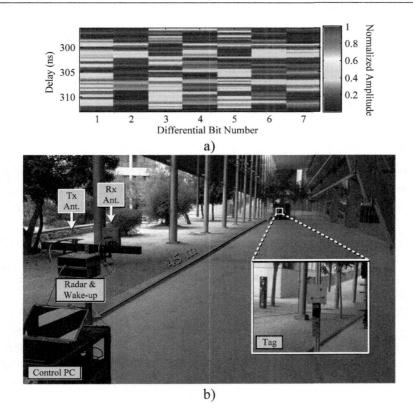

Figure 6.19. *a) Normalized maximum of CWT as a function of time delay for a bit sequence at a tag-to-reader distance of 45 m (outdoor environment). b) Photograph of the measurement setup in the outdoor environment. For a color version of the figure, see www.iste.co.uk/ramos/rfid.zip*

As can be observed from the measurements, the use of cross-polarized antennas helps to reduce both clutter interference and coupling between antennas at the reader and the tag. In the case of the reader, it helps to exploit the dynamic range of the receiver. Commercial dual-port dual-polarization antennas can be used to reduce the reader size and optimize tag-reader alignment [MIC 14]. In the case of the tag, the in-band coupling is reduced by using different polarizations in reception and transmission. Although a good performance is achieved, power consumption could be improved by replacing the commercial MMIC amplifier by a UWB CMOS amplifier.

6.3. Active UWB RFID system based on reflection amplifier

In this case, the tag consisting of a reflection amplifier is connected at the end of a transmission line connected to a UWB antenna. A similar approach has been previously proposed for narrowband RFID transponders [CHA 11], but it has not been explored for UWB RFID systems.

6.3.1. *Introduction*

A scheme of the system is shown in Figure 6.20. Similar to the cross-polarization amplifier in section 6.2, a UHF link is used for the wake-up interrogation. The same time domain radar (section 2.3.2) is also used for the UWB communication. Figure 6.21 shows an image of the tag. The UWB antenna is based on the bow-tie antenna presented in section 2.5.1. The tag size is 63.2 mm width × 103.6 mm height, and it is fabricated on Rogers RO4003C substrate (see Table 2.3).

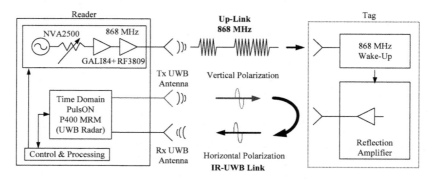

Figure 6.20. *Scheme of the UHF-UWB reader-tag system based on a reflection amplifier*

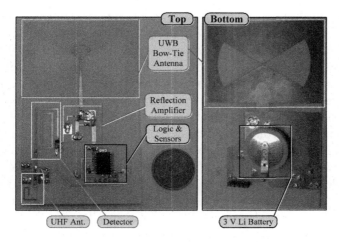

Figure 6.21. *Photograph of the implemented tag*

Considering the circuit theory of section 2.2, the load impedance in this tag corresponds to the input impedance of the one-port reflection amplifier ($\Gamma_{\text{LOAD}} = \Gamma_{\text{IN,AMP}}$), as shown in the scheme of Figure 6.22. When the amplifier is turned off, it presents mismatched load impedance, and $|\Gamma_{\text{LOAD}}|$ is close to 1. Conversely, when the amplifier is turned on, it presents a negative resistance and $|\Gamma_{\text{LOAD}}| > 1$. In this way, the tag mode is modulated. This negative resistance is also used to improve the detection. Usually Structural-to-tag mode ratios are small in time-coded chipless designs, mainly because the structural mode cannot be controlled accurately (see section 2.6). In this design, this poor ratio is improved by using the reflection amplifier gain.

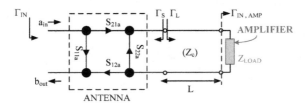

Figure 6.22. *Model for the UWB RFID tag based on reflection amplifier*

6.3.2. *Reflection amplifier design*

A block diagram of a one-port negative resistance amplifier is shown in Figure 6.23. To obtain reflection gain, the reflection coefficient of a one-port network must be greater than 1, $|\Gamma_{\text{IN,AMP}}| > 1$. In this case, the amplifier ought to present negative impedance at its input. The condition for a stable oscillation is $\Gamma_{\text{IN,AMP}} \cdot \Gamma_S > 1$. Here, the reflection gain is conditioned by the return loss of the antenna (which should be higher than the amplifier reflection coefficient). Moreover, in this case the spurious oscillations are damped. The wideband one-port reflection amplifier is designed using a MESFET CFY30 from Infineon. The transistor is biased to $V_{\text{DS}} = 3\ V$ and $I_{\text{DS}} = 6$ mA. This bias voltage is provided by the PIC microcontroller, as with the cross-polarization amplifier from section 6.2. The Maxim MAX4715 switch is used to render the impedance of the amplifier circuit independent from the impedance of the PIC output port. The DC power consumption is 18 mW. A source capacitor and an open-ended stub provide positive feedback to the transistor to generate negative resistance at the operation frequency.

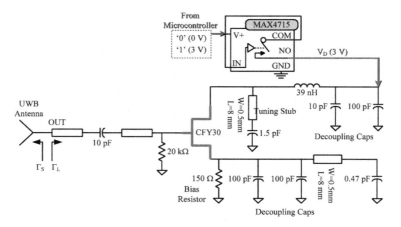

Figure 6.23. *Scheme of the reflection amplifier*

The transistor is biased using the source bias technique. In order to obtain the negative bias point for the V_{GS}, a 150 Ω resistance is connected to the source through a high impedance $\lambda/4$ microstrip transmission line (λ being the wavelength in that medium), and the gate is grounded using a high value resistance (20 kΩ) acting as an RF choke. The 100 pF source decoupling capacitors are required to ensure that there is no RF power loss on the source resistance. A 39 nH RF choke connected to the drain presents a high impedance at RF signal and isolates the drain from the DC bias network. In order to tune the frequency of the peak gain at the center frequency of the reader's UWB pulse generator (around 4.3 GHz), a 1.5 pF capacitor is connected at the end of the stub. A DC block of 10 pF capacitor is connected at the output line. Figure 6.24 compares the measured and simulated reflection coefficients of the amplifier. The amplifier exhibits a negative resistance from 2 to 5 GHz. Its return gain ($|S_{11}|$) at 4.5 GHz is 10.2 dB and is higher than 5 dB from 3.6 to 4.8 GHz.

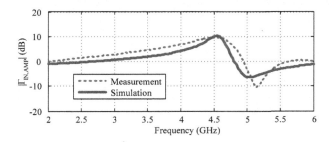

Figure 6.24. *Comparison between measured (dashed line) and simulated (solid line) reflection coefficient of the reflection amplifier*

In order to avoid oscillations due to the tag amplifier, the antenna return loss should be higher than 10 dB in the frequency band where the amplifier presents gain. As shown in the measured return loss of the bow-tie antenna (see section 2.5.1), this is achieved. Since a short straight line is connected to the antenna, the return loss is the same as shown in Figure 2.17(a).

6.3.3. *UWB link budget*

As in section 6.2.3, the link budget of the system is calculated for this amplifier and compared to a passive case. The UWB tag antenna gain considered is 3 dB (which is consistent with the maximum simulated gain of 2.97 dB at 4.3 GHz in section 2.5.1). Figure 6.25 shows the energy per bit to noise spectral density as a function of distance. For a BER lower than 10^{-4}, the theoretical read range would be 11.5 and 7.6 m for an active tag and a passive tag (without the amplifier), respectively. The UHF link budget is identical to the one in section 6.2.3, with a maximum theoretical distance of 45.5 m.

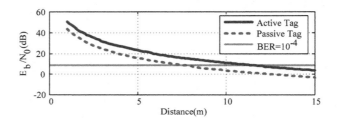

Figure 6.25. *Energy per bit to noise spectral density as a function of distance for an active tag and a passive tag*

6.3.4. *Results*

In order to validate the design, some experiments have been performed. Figure 6.26 shows the normalized magnitude of the CWT of the received signal for the ON state (amplifier biased on) and OFF state (amplifier biased off). The tag-to-reader distance is 1 m and the background has been subtracted from the received signal in order to remove clutter. The first peak corresponds to the structural mode and, obviously, does not depend on the state of the amplifier. On the other hand, the amplitude of the tag mode is noticeably higher in the ON state than in the OFF state due to the return gain of the amplifier. The delay corresponds to the transmission line between the antenna and the amplifier. The ratio of the tag mode peaks between the ON and OFF states corresponds to 2.3 (7.23 dB), which concurs with the peak gain of the amplifier given in Figure 6.24 at 4.3 GHz (center frequency of the radar). It is difficult to distinguish the tag mode from the noise floor for the OFF state whereas it is clearly visible (and also its own multiple reflections every $2L/v$ s) for the ON state.

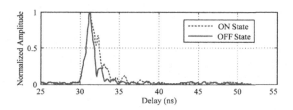

Figure 6.26. *Normalized CWT magnitude for the two tag states at 1 m between the tag and the reader*

Figure 6.27(a) shows the maximum magnitude of the CWT for a bit sequence in an indoor scenario at 10 m, a value which is close to the theoretical limit for a free-space scenario (11.5 m calculated in section 6.3.3). Since the structural-to-tag modes delays do not depend on the tag-to-reader distance, localization algorithms could use the delay information to determine the tag position using multiple readers. Assuming that clutter is stationary, its interference can be removed by subtracting a reference measurement. Figure 7.27(b) shows the differential signal taking the received signal for the first bit as the reference signal. The structural mode is evidently cancelled and the clutter is considerably reduced.

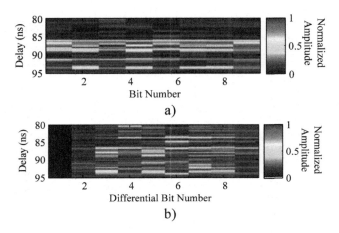

Figure 6.27. *Normalized CWT magnitudes a) and differential signal b) at 10 m. The first bit is the reference for the differential signal. For a color version of the figure, see www.iste.co.uk/ramos/rfid.zip*

6.4. Discussion and comparison between systems

Two alternatives for active time-coded UWB RFID systems have been shown. Both systems increase the read range of time-coded UWB RFID, enabling long-range localization applications in large indoor/outdoor scenarios. The approach based on a commercial amplifier in cross-polarization consumes more power, but provides the largest read range. The reflection amplifier is potentially lower cost, since it is based only on a transistor and passive elements. Regarding tag sizes, the cross-polarization tag, which has two antennas (which cannot be placed very close to each other) is noticeably larger. Table 6.1 summarizes the main parameters of both approaches.

Parameter	Cross-polarization (ABA31563)	Reflection (CFY30)
Supply voltage	3 V	3 V
Supply current	16 mA	6 mA
Gain @ 4.3 GHz	16 dB	10.2 dB
Gain @ 3.1–4.8 GHz	> 15 dB	> 5 dB
Theoretical maximum distance	45.5 m	11.5 m
Measured maximum distance	45 m	10 m
Tag size	Large	Small

Table 6.1. *Comparison between active time-coded UWB RFID systems*

The shorter read range with the reflection amplifier is mainly because it has more of a narrowband behavior. Even though the peak gain around the design frequency is large, it is considerably reduced for frequencies outside the peak. With the ABA31563 amplifier, however, the gain is maintained over a larger band. The reflection amplifier is a lower cost, simpler and lower power alternative, but the read range is not noticeably increased with respect to the semi-passive digital tag (see section 4.3) despite its much higher power consumption.

Figure 6.28 shows a curve with the expected liftetime (in years) for the active tags, the semi-passive digital tag (see section 4.3) and a typical XBee® 802.15.4 module (with a transmission consumption of 45 mA and read range of 30–90 m) [DIG 15]. As expected, the semi-passive digital tag achieves the longest battery duration. However, we can see that the active UWB time-coded RFID tag with cross-polarization could achieve longer battery duration than a Zigbee module, with the added capability of time domain based localization.

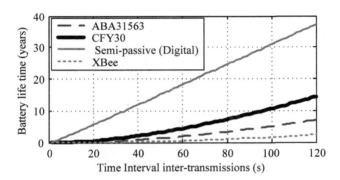

Figure 6.28. *Battery lifetime in years as a function of the time interval between transmissions for the active tags, the semi-passive digital tag, and an XBee module*

Indoor Localization with Smart Floor Based on Time-coded UWB RFID and Ground Penetrating Radar

7.1. Introduction

RFID can be used to locate or guide autonomous entities, such as robots or people, within a defined surface. To this end, tags are scattered in a given space and then an RFID reader is placed on the mobile subject in order to identify the position inside that space, based on a previous tag mapping of the environment [HAN 07, WON 10, JIM 12, OLS 13, KOC 07, PAR 13]. This concept is often referred to as smart floor [GON 13]. Several RFID approaches have been investigated for this application [HAN 07, WON 10, JIM 12, OLS 13, KOC 07, PAR 13, GON 13, TES 10] and low-cost passive tags are preferred [GON 13]. For instance, 13.56 MHz tags are used in [GON 13, TES 10], while UHF passive tags are used in [PAR 13].

The time-coded UWB RFID systems discussed in this book can be an alternative for smart floors and indoor localization applications. Since they are based in the time domain, easy localization is possible using the delay information. In addition, as explained in section 2.6.3, time-coded UWB RFID tags are not heavily affected by the material they are attached to. Therefore, they can be embedded in different types of floor.

This chapter performs a study on the feasibility of using time-coded UWB RFID tags and its combination with PR techniques for indoor mapping, localization and guidance. To this end, three types of tags are considered as follows:

1) passive reflectors that are made up of metallic strips;

2) chipless time-coded UWB tags, as discussed in Chapter 2;

3) semi-passive time-coded UWB tags (see section 4.3).

Since the UWB radar used as reader transmits a pulse and the tag responds by backscattering, the ToA can be measured without the need of any synchronization system implemented in the tag. This chapter is organized as follows:

– section 7.2 shows the alternatives for smart floor design;

– section 7.3 presents the results obtained for the passive reflectors, and the chipless and semi-passive time-coded UWB tags;

– finally, section 7.4 draws conclusions.

7.2. Smart floor design alternatives

The tag-reader system is shown in Figure 7.1(a). The reader is based on the bistatic UWB radar Time Domain PulsON P400 MRM (see section 2.3.2). It is connected to two UWB antennas (Tx and Rx) that illuminate the smart floor, which is shown in Figure 7.1(b). The smart floor consists of ceramic tiles with a total size of 1 × 2 m. It has been constructed on top of the lab floor, with a 3 cm separation to embed the tags. The tags are buried under the smart floor using top and bottom spacers (foam). The system (reader + antennas) is supported by a mobile platform that sweeps the floor.

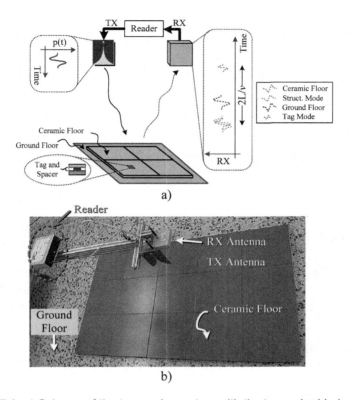

Figure 7.1. *a) Scheme of the tag reader system with the tag embedded under the floor, b) photograph of the smart floor and the measurement equipment. For a color version of the figure, see www.iste.co.uk/ramos/rfid.zip*

Three different tags are studied and briefly introduced here. Diagrams of the three tags are shown in Figure 7.2. The first approach is to use tags based on passive reflectors (metal strips). The information can be coded in the spacing between the reflectors and in the number of reflectors. The second approach is based on chipless time-coded tags. These tags consist of a UWB antenna connected to an open-ended delay line, as presented in section 2.2. Therefore, different delays can be used to encode information. The last approach is based on a semi-passive time-coded tag. Here, the delay line is connected to a load that is modulated to transmit the information (see section 4.3).

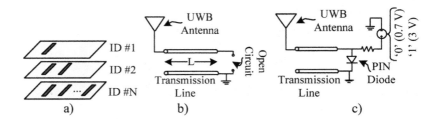

Figure 7.2. *a) Examples of tags based on passive reflectors, b) block diagram of time-coded tag and c) block diagram of semi-passive time-coded tag*

The first and second approaches are very simple, and their capacity to store information is limited so can effectively be used for indoor guidance (for instance with tag IDs that correspond to the orders of "straight on", "turn left" or "turn right"). The third approach permits the storage of large amounts of information (since a microcontroller is embedded). As a result, it can be used not only for guidance but also to identify places or provide information on objects in indoor spaces.

7.3. Results

7.3.1. *Smart floor based on passive reflectors*

GPR is a well-established technology employed to detect buried objects (pipes, archeological remains, etc.) or boundaries between different dielectric constants [DAN 05]. Inspired by GPR technology, here a smart floor for indoor localization and guidance is shown based on tags that consist of several passive reflectors (topology shown in Figure 7.2(a)). Information can be coded in the number of passive reflectors and the distance between them. On the limit, a tag may consist of a simple metallic strip. Here, the UWB radar is used as a GPR. The depth range of GPR is limited by ground loss, the transmitted center frequency and the radiated power. Penetration depth decreases when loss and frequency increase. Conversely, resolution increases with frequency.

There are some important differences between this application and standard GPR technology as follows:

– first, the center frequency of the UWB radar (about 4.3 GHz) is higher than GPR (typically between 0.9 and 2 GHz for detection of buried pipes). As a result, the resolution is higher. This is essential to detect a small tag buried just under the floor surface, since floor and tag reflections are very close in time. Normally, GPR is used to detect objects (e.g. pipes) buried at depths of about 0.5–1 m below the floor level;

– second, the spectrum of the UWB radar complies with UWB FCC mask, which is more restrictive than the regulation of GPR frequency bands, and as a result less power is transmitted;

– third, GPR antennas are generally in close proximity to the surface. The strong reflection due to the floor surface can be deleted using a timewindow, as explained; generally, the buried objects are far from the surface. Here, the tag is very close to the floor and the tag response cannot be easily filtered. In order to avoid the blind distance effect, the antennas are airlaunched allowing filtering of the signal coupled between TX and RX antennas;

– fourth, small, portable and low-cost CMOS UWB radars (see section 2.3.2) are commercially available and suitable for this application when compared with heavy and expensive GPR equipment.

Simulations are achieved using the finite-difference time-domain method [GPR 14]. The simulated scenario consists of a floor of 1 cm thick ceramic tiles ($\varepsilon_r = 6$). A perfect electrical conductor (PEC) sheet is buried between the floor and the ground (which are separated by a distance of 5 cm). The ground is simulated with a dielectric permittivity close to dry sand ($\varepsilon_r = 3$). A spacer of 1 cm is considered in order to simulate the experimental test bed scenario. The Tx and Rx antennas are separated with 10 cm between them, and the radar scans by a distance of 1 m.

Figure 7.3 shows the raw (unprocessed) data for the cases of two individual PEC sheets 10 and 20 cm wide. Figure 7.4 shows the results after clutter removal by applying background subtraction. Clutter is mainly due to antenna coupling and reflection at the floor surface. It is removed by using the average of the image in cross-range [LAZ 14]. This background subtraction can be done dynamically using a moving averaging filter, as described in [LAZ 14]. After removing clutter, the position of the reflector can be easily obtained. The typical hyperbola shape, as in GPR systems, can be observed [GPR 14]. The maximum is located at the center of the reflector. The depth of the reflector can be estimated from the delay at the maximum point and the propagation velocity of the medium from the aperture of the hyperbola.

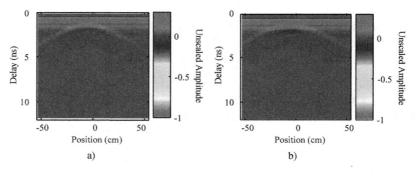

Figure 7.3. *Simulated raw data for a single PEC sheet of width a) 10 cm and b) 20 cm. For a color version of the figure, see www.iste.co.uk/ramos/rfid.zip*

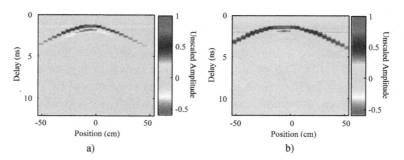

Figure 7.4. *Simulated data for a single PEC sheet of width a) 10 cm and b) 20 cm after background subtraction. For a color version of the figure, see www.iste.co.uk/ramos/rfid.zip*

Figure 7.5 shows the simulated raw data for two sheets 10 cm wide separated by 30 cm and 50 cm, respectively. Figure 7.6 shows the same simulation after background subtraction. It can be observed that a ghost reflection appears between the two sheets and at a larger delay due to the coupling between the two reflections. The intensity of this ghost reflection is increased when the spacing between the reflectors is reduced.

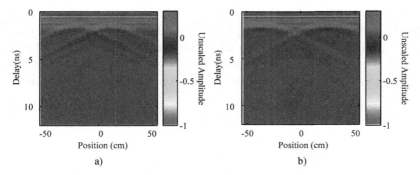

Figure 7.5. *Simulated raw data for two PEC sheets of width 10 cm separated by a) 30 cm and b) 50 cm. For a color version of the figure, see www.iste.co.uk/ramos/rfid.zip*

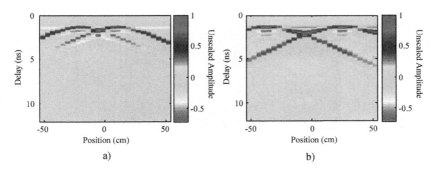

Figure 7.6. *Simulated data for two PEC sheets of width 10 cm separated by a) 30 cm and b) 50 cm after background subtraction. For a color version of the figure, see www.iste.co.uk/ramos/rfid.zip*

Concerning measurements, as an example three tag IDs (as shown in Figure 7.2(a)) are measured by considering one, two or three metallic sheets with different widths (15/10/20 cm), separated by 25 cm. Figure 7.7 shows the raw measured data. It can be observed that the raw data are enough to detect the position and the hyperbolas corresponding to the metal sheets. This is an interesting result since it means that no background subtraction is required for this application and as a result, inhomogeneities in the floor are supported.

7.3.2. *Smart floor based on chipless time-coded UWB RFID tags*

It is shown in Figure 7.1(a) that a long delay line connected to the UWB antenna is required to separate the tag mode from the reflections at the ground. To this end, two tags have been emulated using a UWB antenna and coaxial delay lines connected to an RF switch. In this manner, the minimum delay required to separate the tag mode can be studied.

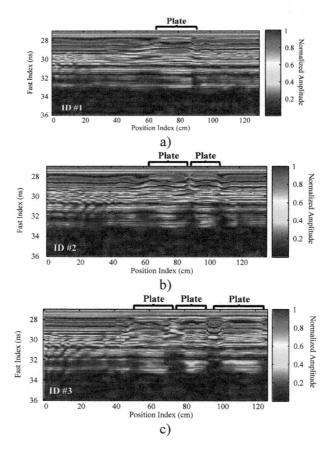

Figure 7.7. *Raw measured data for three tag IDs based on metal sheet reflectors; a) one reflector, b) two reflectors and c) three reflectors. For a color version of the figure, see www.iste.co.uk/ramos/rfid.zip*

Figure 7.8 shows the two time-coded chipless UWB RFID tags read under the floor structure. The tag modes are marked with red arrows. Background subtraction is applied for clutter removal. As can be observed, depending on the tag ID (length of the delay line), the tag mode appears at a different time. Since the tag has a small RCS compared to a metal sheet, it is difficult to detect the hyperbola shape in this case, and the estimation of the position of the tag is more inaccurate. However, for identification purposes, the tag can be read. A combination of this approach with the passive reflectors (see section 7.3.1), which could be realized by simply increasing the RCS of the tag by adding a metal strip, would lead to the possibility of better tag detection (structural mode).

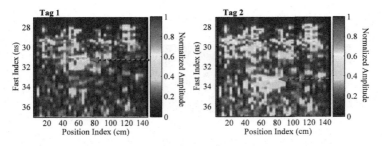

Figure 7.8. *Measured chipless UWB tags under the ceramic floor for two tag IDs. For a color version of the figure, see www.iste.co.uk/ramos/rfid.zip*

Figure 7.9 shows the cuts at 59.34 cm for Figure 7.8. This distance of 59.34 cm corresponds to the maximum position index for both tags from the beginning of the smart floor. The difference between the tag mode delays can be clearly observed. In addition, the peaks corresponding to the ceramic floor, ground floor and structural modes are also observed.

In order to detect the hyperbola shape with a small tag, additional techniques to reduce clutter have to be taken into account. For instance, from the measurement of Figure 7.8, the differential signal between the two tags is calculated (similarly to as in section 4.3.4) and shown in Figure 7.10. The hyperbola shape can now be observed at the delays corresponding to the two tag modes. The structural mode, clutter, and peaks associated to the floor are greatly reduced.

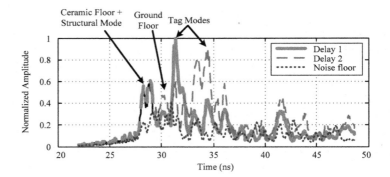

Figure 7.9. *Cut at 59.34 cm of the measured chipless UWB tags under the ceramic floor for two tag IDs . For a color version of the figure, see www.iste.co.uk/ramos/rfid.zip*

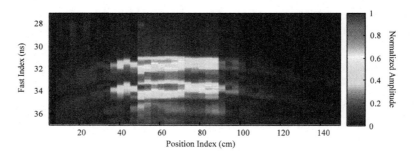

Figure 7.10. *Differential signal between the two chipless tag IDs. For a color version of the figure, see www.iste.co.uk/ramos/rfid.zip*

7.3.3. *Smart floor based on semi-passive time-coded UWB RFID tags*

If high precision is required, a semi-passive approach can be considered. Even though a battery-based tag (see section 4.3) has been considered for this section, a tag powered by power-scavenging techniques could be used. This would make possible the integration of the tag in embedded structures for long-term applications. The use of semi-passive approaches permits two tag states and performs differential measurements. Figure 7.11 shows the differential signal for a semi-passive tag (see section 4.3). The tag is measured two times at different positions. The hyperbola shape is perfectly detected, and the clutter and reflections from the floor are greatly reduced. It is important to note that for this application, where the tag is close to the floor surface, it is possible to send the wake-up signal at 2.45 GHz, fulfilling all regulations in order to operate the tag.

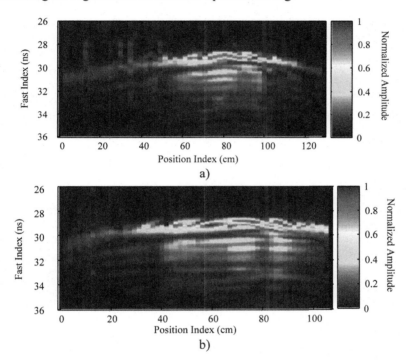

Figure 7.11. *Differential signal for the semi-passive time-coded UWB tag; a) position 1 and b) position 2. For a color version of the figure, see www.iste.co.uk/ramos/rfid.zip*

7.4. Conclusions

This chapter has presented a study on the feasibility of using time-coded UWB RFID tags for indoor mapping, localization and guidance. Detection techniques based on GPR applications have been used and combined with time-coded UWB RFID tags. Detection of buried tags based on passive reflectors, chipless time-coded RFID and semi-passive time-coded RFID have been presented. The passive reflectors and the chipless time-coded UWB tags provide a low-cost and simple solution for smart floors for tracking and guidance. However, they are less robust to noise and they have a lower precision than semi-passive tags. Semi-passive tags are a more expensive and complex solution, but with higher reliability and precision and the possibility to store information.

Bibliography

[ABR 04] ABRAHAM J.K., PHILIP B., WITCHURCH A. *et al.* "A compact wireless gas sensor using a carbon nanotube/PMMA thin film chemiresistor", *Smart Materials and Structures*, vol. 13, no. 5, pp. 1045–1049, 2004.

[AGI 04] AGILENT TECHNOLOGIES, De-embedding and Embedding S-Parameter Networks Using a Vector Network Analyzer, Application Note 1364-1, 2004.

[AGI 12] AGILENT TECHNOLOGIES, Agilent time domain analysis using a network analyzer, Application Note 1287-12, 2012.

[ALI 14] ALIEN TECHNOLOGY, "Higgs(R) 4 RFID IC", available at http://www.alientechnology.com/wp-content/uploads/ALC-370-SOT%20 Higgs4%20SOT%202014-10-17.pdf, 2014.

[ALO 10] ALONG K., CHENRUI Z., LUO Z. *et al.*, "SAW RFID enabled multi-functional sensors for food safety applications", *IEEE International Conference on RFID-Technology and Applications*, Guangzhou, China, June 2010.

[AME 14] AMERICAN SOCIETY FOR TESTING AND MATERIALS (ASTM), Terrestrial Reference Spectra for Photovoltaic Performance Evaluation. Available at http://rredc.nrel.gov/solar/spectra/am1.5/, 2014.

[ANG 06] ANGELOPOULOS S., ANASTOPOULOS A.Z., KAKLAMANI D.I. *et al.*, "Circular and elliptical CPW-fed slot and microstrip-fed antennas for ultra wideband applications", *IEEE Antennas and Wireless Propagation Letters*, vol. 5, pp. 294–297, 2006.

[AST 09] ASTRIN A.W., LI H.B., KOHNO R., "Standardization for body area networks", *IEICE Transactions on Communication*, vol. E92-B, no. 2, pp. 366–372, 2009.

[ATM 12] ATMEL, "8-bit AVR Microcontroller with 4K Bytes In-System Programmable Flash and Boost Converter", ATtiny43U Datasheet, available at http://www.atmel.com/Images/8048s.pdf, 2012.

[AVA 14] AVAGO TECHNOLOGIES, "HSMS-285x series surface mount zero bias Schottky detector diodes", available at http://www.avagotech.com/docs/AV02-1377EN, 2014.

[BAG 09] BAGHAEI-NEJAD M., MENDOZA D. S., ZOU Z. *et al.*, "A remote-powered RFID Tag with 10Mb/s UWB uplink and -18.5 dBm sensitivity UHF downlink in 0.18 µm CMOS", *Digest of Technical Papers, ISSCC*, San Francisco, California, USA, 8–12 February 2009.

[BAL 09a] BALBIN I., KARMAKAR N.C., "Phase-encoded chipless RFID transponder for large-scale low-cost applications", *IEEE Microwave and Wireless Components Letters*, vol. 19, no. 8, pp. 509–511, 2009.

[BAL 09b] BALBIN I., KARMAKAR N.C., "Novel chipless RFID tag for conveyor belt tracking using multi-resonant dipole antenna", *Proceedings of the 39th European Microwave Conference*, Rome, Italy, pp. 1109–1112, 29 September – 01 October 2009.

[BER 91] BERNARD P.A., GAUTRAY J.M., "Measurement of dielectric constant using a microstrip ring resonator", *IEEE Transactions on Microwave Theory and Techniques*, vol. 39, no. 3, pp. 592–595, 1991.

[BHA 10] BHATTACHARYYA R., FLOERKEMEIER C., SARMA S., "RFID tag antenna based temperature sensing", IEEE *Proceedings of IEEE International Conference on RFID*, Orlando, USA, pp. 8–15, 14–16 April 2010.

[BRU 84] BRUNFELDT D.R., ULABY F.T., "Active reflector for radar calibration", *IEEE Transactions on Geoscience and Remote Sensing*, vol. (GE-22), no. 2, pp.165–169, 1984.

[CAI 11] CAIZZONE S., OCCHIUZZI C., MARROCCO G., "Multi-chip RFID antenna integrating shape-memory alloys for detection of thermal thresholds", *IEEE Transactions on Antennas and Propagation*, vol. 59, no. 7, pp. 2488–2494, 2011.

[CAR 07] GOMES H.C., CARVALHO N.B., "The use of inter modulation distortion for the design of passive RFID", *Proceedings of the European Microwave Conference*, Munich, Germany, pp. 1656–1659, 2007.

[CHA 11] CHAN P., FUSCO V., "Bi-Static 5.8 GHz RFID range enhancement using retro directive techniques", *Proceedings of the 41st European Microwave Conference*, Manchester, UK, pp. 976–979, October 2011.

[CHE 09] CHEN Z.N., QING X., CHUNG H.L., "A universal UHF RFID reader antenna", *IEEE Transactions on Microwave Theory and Techniques*, vol. 57, no. 5, pp.1275–1282, 2009.

[CHE 10] CHE W., MENG D., XANG X. *et al.*, "A semi-passive UHF RFID tag with on-chip temperature sensor", *IEEE Custom Integrated Circuits Conference*, San Jose, California, USA, pp. 1–4, 2010.

[CHO 05] CHO N., SONG S-J., KIM S. *et al.*, "A 5.1-μW UHF RFID tag chip integrated with sensors for wireless environmental monitoring", *Proceedings of the 31st European Solid-State Circuits Conference*, Grenoble, France, pp. 279–282, 2005.

[COL 04] COLLINS J., "Alien cuts tag price", RFID Journal, available at http://www.rfidjournal.com/articles/view?857, April 2004.

[COL 06] COLLINS J., "RFID fibers for secure applications", RFID Journal, available at http://www.rfidjournal.com/article/view?845, 2006.

[COL 13] COLLADO A., GEORGIADIS A., "Conformal hybrid solar and electromagnetic (EM) energy harvesting rectenna", *IEEE Transactions on Circuits and Systems I: Regular Papers*, vol. 60, no. 8, pp. 2225–2334, 2013.

[COL 69] COLLIN R.E., ZUCKER F.J., *The Receiving Antenna, Antenna Theory: Part 1*, McGraw-Hill, New-York, 1969.

[COW 66] COWLEY A.M., SORENSEN H.O., "Quantitative comparison of solid-state microwave detectors", *IEEE Transactions on Microwave Theory and Techniques*, vol. 14, no. 12, pp. 588–602, 196.

[D'ER 12] D'ERRICO R. BOTTAZZI M., NATALI F. *et al.*, "An UWB-UHF semi-passive RFID system for localization and tracking applications", *IEEE International Conference on RFID Technology and Applications*, Nice, France, pp.18–23, 2012.

[DAN 05] DANIELS D.J., *Ground Penetrating Radar*, John Wiley & Sons, New York, 2005.

[DAR 04] DARDARI D., "Pseudo-random active UWB reflectors for accurate ranging", *IEEE Communications Letters*, vol. 8, no. 10, pp. 608–610, 2004.

[DAR 08] DARDARI D., D'ERRICO R., "Passive ultrawide bandwidth RFID", *IEEE Global Telecommunications Conference (GLOBECOM)*, New Orleans, pp. 1–6, 2008.

[DAS 06] DAS R., HARROP P., Chipless RFID Forecasts, Technologies & Players 2006-2016, IDTech Ex report, UK, 2006.

[DE 13] DE DONNO D., CATARINUCCI L., TARRICONE L., "Enabling self-powered autonomous wireless sensors with new-generation I2C-RFID chips", *IEEE MTT-S International Microwave Symposium Digest (MTT)*, Seattle, WA, June 2013.

[DEM 13] DEMENTYEV A., HODGES S., TAYLOR S. *et al.*, "Power consumption analysis of bluetooth low energy, ZigBee and ANT sensor nodes in a cyclic sleep scenario", *IEEE International Wireless Symposium*, Beijing, pp. 1–4, 2013.

[DEN 14] DENSO W., "What is a QR code?", http://www.qrcode.com/en/about/, 2014.

[DI 06] DI BENEDETTO M.G., KAISER T., MOLISH A.F. *et al* (eds.)., *UWB Communications Systems: A Comprehensive Overview*, Hindawi Publishing Corporation, 2006.

[DIG 15] DIGI INTERNATIONAL INC, *XBee® 802.15.4 – Device Connectivity Using Multiport Wireless Networks*, Digi International Inc, available at http://www.   digi.com/products/xbee-rf-solutions/modules/xbee-series1-module, 2015.

[DOU 10] HU S., ZHOU Y., LAW C.L. *et al.*, "Study of a uniplanar monopole antenna for passive chipless UWB-RFID localization system", *IEEE Transactions on Antennas and Propagation*, vol. 58, no. 2, pp. 271–278, 2010.

[DOW 09] DOWLING J., TENTZERIS M.M., "'Smart house' and 'smart-energy' applications of low-power RFID-based wireless sensors", *IEEE Asia Microwave Conference*, Singapore, pp. 2412–2415, December 2009.

[DOW 09] DOWLING J., TENTZERIS M.M., BECKET N., "RFID-enabled temperature sensing devices: a major step forward for energy efficiency in home and industrial applications?", *IEEE MTT-S International Microwave Workshop on Wireless Sensing, Local Positioning, and RFID*, Cavtat, Croatia, pp. 1–4, 24–25 September 2009.

[DYN 15] DYNASTREAM INNOVATIONS INC., "This is ANT", http://www.this isant.com/, 2015.

[ETS 08] ETSI EN 302 065 V1.1.1 (2008-02), Electromagnetic compatibility and radio spectrum matters (ERM); Ultra-wideband (UWB) technologies for communication purposes; Harmonized EN covering the essential requirements of article 3.2 of the R&TTE directive, available at http://www.etsi.org/deliver/etsi_en/302000_302099/30206501/01.03.01_20/en_30206501v010301a.pdf, February 2008.

[EXT 10] EXTRONICS OMNI ID, "RFID Tag Comparison Guide", http://www.extronics.com/media/234005/rfid_tag_comparison_broadband_whitepaper.pdf, 2010.

[FCC 03] FCC NOTICE OF PROPOSED RULE MAKING, Revision of Part 15 of the Commision's Rules Regarding Ultra-wideband Transmission Systems, ET-Docket 98-153, FCC, 2003.

[FER 11] FERNANDES R.D., BOAVENTURA A.S., CARVALHO N.B. *et al.*, "Increasing the range of wireless passive sensor nodes using multisines", *IEEE International Conference on RFID-Technologies and Applications*, Barcelona, pp. 549–553, September 2011.

[FIN 10] FINKENZELLER K., *RFID Handbook: Fundamentals and Applications in Contactless Smart Cards, Radio Frequency Identification and Near-Field Communication*, Wiley-Blackwell, 2010.

[FLE 02] FLETCHER R.R., *Low-Cost Electromagnetic Tagging: Design and Implementation*, Massachusetts Institute of Technology, USA, 2002.

[FON 02] FONSECA M.A., ENGLISH J.M., ARX M.V. *et al.*, "Wireless micromachined ceramic pressure sensor for high-temperature applications", *IEEE Journal of Microelectromechanical Systems*, vol. 11, no. 4, pp. 337–343, 2002.

[FON 04] FONTANA R. J., "Recent system applications of short-pulse ultra-wideband (UWB) technology", *IEEE Transactions on Microwave Theory and Techniques*, vol. 52, no. 9, pp. 2087–2104, 2004.

[FU 04] FU Y., DU H., HUANG W. *et al.*, "TiNi-based thin films in MEMS applications: a review", *Sensors and Actuators A*, vol. 112, pp. 395–408, 2004.

[GAO 10] GAO J., SIDEN J., NILSSON H-K., "Printed temperature sensors for passive RFID tags", *Proceedings of Progress in Electromagnetics Research Symposium*, Xi'an, China, pp. 845–848, March 2010.

[GEO 10] GEORGIADIS A., ANDIA VERA G., COLLADO A., "Rectenna design and optimization using reciprocity theory and harmonic balance analysis for electromagnetic (EM) energy harvesting", *IEEE Antennas and Wireless Propagation Letters*, vol. 9, pp. 444–446, 2010.

[GEO 12] GEORGIADIS A., COLLADO A., KIM S. *et al.*, "UHF solar powered active oscillator antenna on low cost flexible substrate for wireless identification applications", *IEEE MTT-S International Microwave Symposium Digest*, Montreal QC, Canada, pp. 1–3, June 2012.

[GEO 14] GEOZONDAS, "GZ6EVK optional 31106", available at http://www.geozondas.com/main_page.php?pusl=5, 2014.

[GIR 12] GIRBAU D., LAZARO A., VILLARINO R., "Passive wireless permittivity sensor based on frequency-coded chipless RFID tags", *IEEE MTT-S International Microwave Symposium Digest*, Montreal QC, Canada, pp. 1–3, June 2012.

[GON 13] GONALVES R., REIS J., SANTANA E. *et al.*, "Smart floor: indoor navigation based on RFID", *IEEE Wireless Power Transfer*, Perugia, Italy, pp. 103–106, May 2013.

[GON 14] GONCALVES R., RIMA S., MAGUETA R. *et al.*, "RFID tags on cork stoppers for bottle identification", *IEEE MTT-S International Microwave Symposium*, Florida, USA, pp. 1–4, June 2014.

[GPR 14] GPR MAX, "Gpr MAX V2.0", available at http://www.gprmax.com/, 2014.

[GRE 66] GREEN R.B., "Relationships between antennas as scatterers and radiators", *IEEE Transactions on Antennas and Propagation*, vol. 14, no. 1, pp. 17–21, 1966.

[GS1 14] GS1, "Regulatory status for using RFID in the EPC Gen 2 band (860 to 960 MHz) of the UHF spectrum", available at http://www.gs1.org/docs/epcglobal/UHF_Regulations.pdf, 2014.

[GUN 09] GUNGOR V.C., HANCKE G.P., "Industrial wireless sensor networks: challenges, design principles, and technical approaches", *IEEE Transactions on Industrial Electronics*, vol. 56, no. 10, pp. 4258–4265, 2009.

[HAF 13] HAFAIEDH I., ELLEUCH W., CLEMENT P. *et al.*, "Multi-walled carbon nanotubes for volatile organic compound detection", *Sensors and Actuators B: Chemical*, vol. 182, pp. 344–350, 2013.

[HAN 07] HAN S., LIM H., LEE J., "An efficient localization scheme for a differential-driving mobile robot based on RFID system", *IEEE Transactions on Industrial Electronics*, vol. 54, no. 6, pp. 3362–3369, 2007.

[HAN 89] HANSEN R.C., "Scattering from conjugate-matched antennas", *Proceedings of the IEEE*, vol. 77, no. 5, pp. 659–662, 1989.

[HAR 02] HARTMANN C.S., "A global SAW ID tag with large data capacity", *Proceedings of IEEE Ultrasonics Symposium*, Munich, Germany, vol. 1, pp. 65–69, 2002.

[HAR 14] HARROP P., DAS R., HOLLAND G., *Near Field Communication (NFC) 2014-2024 – Mobile Phone and Other NFC: Market Forecasts, Technology, Players*, ID Tech Ex, Cambridge UK, 2014.

[HEL 05] HELAL S., MANN W., EL-ZABADANI H. *et al.*, "The gator tech smart house: a programmable pervasive space", *Computer*, vol. 38, no. 3, pp. 50–60, 2005.

[HER 14] HEMOUR S., WU K., "Radio-frequency rectifier for electromagnetic energy harvesting: development path and future outlook", *Proceedings of the IEEE*, vol. 102, no. 11, pp. 1667–1691, 2014.

[HEW 82] HEWLETT-PACKARD, All Schottky diodes are zero bias detectors, Application Note 988, 1982.

[HU 07] HU S., LAW C.L., DOU W., "Petaloid antenna for passive UWB-RFID tags", *Electronics Letters*, vol. 43, no. 22, pp. 1174–1176, 2007.

[HU 08] HU S., LAW C.L., DOU W., "A balloon-shaped monopole antenna for passive UWB-RFID tag applications", *IEEE Antennas and Wireless Propagation Letters*, vol. 7, pp. 366–368, 2008.

[HU 10] HU S., ZHOU Y., LAW C.L. *et al.*, "Study of a uniplanar monopole antenna for passive chipless UWB-RFID localization system", *IEEE Transactions on Antennas and Propagation*, vol. 58, no. 2, pp. 271–278, 2010.

[HUA 10] HUANG W.M., DING Z., WANG C.C. *et al.*, "Shape memory materials", *Materials Today*, vol. 13, no. 7–8, pp. 54–61, 2010.

[HUI 08] HUI X., BANGYU L., SHAOBO X. *et al.*, "The measurement of dielectric constant of the concrete using single-frequency CW radar", *IEEE 1st International Conference on Intelligent Networks and Intelligent Systems*, Wuhan, China, pp. 588–591, 2008.

[IMP 14] IMPINJ INC., "Monza R6 RFID Tag Chip", available at http://www.impinj.com/products/tag-chips/monza-r6/, 2014.

[ION 06] IONESCU R., ESPINOSA E.H., SOTTER E. *et al.*, "Oxygen functionalisation of MWNT and their use as gas sensitive thick-film layers", *Sensors and Actuators B*, vol. 113, pp.36–46, 2006.

[JAB 10] JABBAR H., SONG Y.S., JEONG T.T., "RF energy harvesting system and circuits for charging of mobile devices", *IEEE Transactions on Consumer Electronics*, vol. 56, no. 1, pp. 247–253, 2010.

[JAL 05] JALALY I., ROBERTSON I.D., "RF bar codes using multiple frequency bands", *Proceedings of IEEE MTT-S International Microwave Symposium Digest*, pp. 4–7, 2005.

[JAT 09] JATLAOUI M.M., CHEBILA F., GMATI I. *et al.*, "New electromagnetic transduction micro-sensor concept for passive wireless pressure monitoring application", *15th International Conference on Solid-State Sensors, Actuators and Microsystems*, Denver, USA, June 2009.

[JIM 12] JIMENEZ A.R., SECO F., PRIETO J.C. *et al.*, "Accurate pedestrian indoor navigation by tightly coupling foot-mounted IMU and RFID measurements", *IEEE Transactions on Instrumentation and Measurement*, vol. 61, no. 1, pp. 178–189, 2012.

[JOH 82] JOHNSON J.H., CHOI W., MOORE R.L., "Precision experimental characterization of the scattering and radiation properties of antennas", *IEEE Transactions on Antennas and Propagation*, vol. 30, pp. 108–112, 1982.

[JON 07] JONES K.C., Invisible RFID Ink Safe for Cattle and People, Company Says, Information Week, United States, 2007.

[JOS 04] JOSE S., Design of RF CMOS power amplifier for UWB applications, Masters Thesis, Virginia Polytechnic Institute and State University, 2004.

[KAI 94] KAISER G., *A Friendly Guide to Wavelets*, Birkhauser, Boston, 1994.

[KAR 10] KARMAKAR N.C., PRERADOVIC S., "Chipless RFID: bar code of the future", *IEEE Microwave Magazine*, vol. 11, no. 7, pp. 87–97, 2010.

[KEY 14] KEYSIGHT TECHNOLOGIES, "E8364C PNA microwave network analyzer", http://www.keysight.com/en/pd-1350015-pn-E8364C/pna-microwave-network-analyzer?&cc=ES&lc=eng, 2014.

[KIM 12] KIM S., GEORGIADIS A., COLLADO A. *et al.*, "An inkjet-printed solar-powered wireless beacon on paper for identification and wireless power transmission applications", *IEEE Transactions on Microwave Theory and Techniques*, vol. 60, no. 12, pp. 4178–4186, 2012.

[KIM 13] KIM S., MARIOTTI C., ALIMENTI F. *et al.*, "No battery required: perpetual RFID-enabled wireless sensors for cognitive intelligence applications", *IEEE Microwave Magazine*, vol. 14, no. 5, pp. 66–77, 2013.

[KOC 06] KOCER F., FLYNN M.P., "An RF-powered, wireless CMOS temperature sensor", *IEEE Sensors Journal*, vol. 6, no. 3, pp. 557–564, 2006.

[KOC 07] KOCH J., WETTACH J., BLOCH E. *et al.*, "Indoor localisation of humans, objects, and mobile robots with RFID infrastructure", *7th International Conference on Hybrid Intelligent Systems*, Germany, pp. 271–276, September 2007.

[KOR 10] KORTUEM G., KAWSAR F., FITTON D. *et al.*, "Smart objects as building blocks for the Internet of things", *IEEE Internet Computing*, vol. 14, no. 1, pp. 44–51, 2010.

[KRI 10] KRISHNAN S., PILLAI V., WENJIANG W., "UWB-IR active reflector for high precision ranging and positioning applications", *IEEE International Conference on Communication Systems*, Singapore, pp. 14–18, 2010.

[LAK 11] LAKAFOSIS V., XIAOHUA Y., TAORAN L. *et al.*, "Wireless sensing with smart skins", *Proceedings of the IEEE Sensors*, Limerick, Ireland, pp. 623–626, October 2011.

[LAZ 09] LAZARO A., GIRBAU D., VILLARINO R., "Wavelet-based breast tumor localization technique of microwave imaging using UWB", *Progress in Electromagnetic Research*, vol. 94, pp. 264–280, 2009.

[LAZ 09a] LAZARO A., GIRBAU D., VILLARINO R., "Effects of interferences in UHF RFID Systems", *Progress in Electromagnetics Research*, vol. 98, pp. 435–443, 2009.

[LAZ 09b] LAZARO A., GIRBAU D., SALINAS D., "Radio link budgets for UHF RFID on multipath environments", *IEEE Transactions on Antennas and Propagation*, vol. 57, no. 4, pp. 1241–1251, 2009.

[LAZ 11a] LAZARO A., VILLARINO R., GIRBAU D., "Design of tapered slot Vivaldi antenna for UWB breast cancer detection", *Microwave and Optical Technology Letters*, vol. 53, no. 3, pp. 639–643, 2011.

[LAZ 11b] LAZARO A., RAMOS A., GIRBAU D. *et al.*, "Chipless UWB RFID tag detection using continuous wavelet transform", *IEEE Antennas and Wireless Propagation Letters*, vol. 10, pp. 520–523, 2011.

[LAZ 14] LAZARO A., GIRBAU D., VILLARINO R., "Techniques for clutter suppression in the presence of body movements during the detection of respiratory activity through UWB radars", *Sensors*, pp. 2595–2618, 2014.

[LEG 10] LEGHRIB R., FELTEN A., DEMOISSON F. *et al.*, "Room-temperature, selective detection of benzene at trace levels using plasma-treated metal-decorated multiwalled carbon nanotubes", *Carbon*, vol. 48, no. 12, pp. 3477–3484, 2010.

[LIU 03] LIU Y., FU D.M., GONG S.X., "A novel model for analyzing the radar cross section of microstrip antenna", *Journal of Electromagnetic Waves and Applications*, vol. 17, pp. 1301–1310, 2003.

[LIU 08] LIU H., BOLIC M., NAYAK A. *et al.*, "Taxonomy and challenges of the integration of RFID and wireless sensor networks", *IEEE Network*, vol. 22, no. 6, pp. 26–35, 2008.

[LOE 14] LOEB W., 10 Reasons Why Alibaba Blows Away Amazon and eBay, Forbes, USA, 2014.

[LOR 11] LORENZO J., GIRBAU D., LAZARO A., *et al.*, "Read range reduction in UHF RFID due to antenna detuning and gain penalty", *Microwave and Optical Technology Letters*, vol. 53, no. 1, pp. 144–148, 2011.

[LOY 14] LOY M., KARINGATTIL R., ISM-Band and Short Range Device Regulatory Compliance Overview, Application Report SWRA048, Texas Instruments, available at http://www.ti.com/ lit/an/swra048/swra048.pdf, 2014.

[LU 10] LU Y., HUANG Y., CHATTA H.T. *et al.*, "An elliptical UWB monopole antenna with reduced ground plane effects", *International Workshop on Antenna Technology*, Lisbon, pp. 1–4, 2010.

[LU 11] LU Y., HUANG Y., CHATTA H.T. *et al.*, "Reducing ground-plane effects on UWB monopole antennas", *IEEE Antennas and Wireless Propagation Letters*, vol. 10, pp. 147–150, 2011.

[LYN 07] LYNCH J.P., "An overview of wireless structural health monitoring for civil structures", *Philosophical Transactions of the Royal Society A*, vol. 365, no. 1851, pp. 345–372, 2007.

[MAG 09] MAGUIRE Y., "An optimal Q-algorithm for the ISO 18000-6C RFID protocol", *IEEE Transactions on Automation Science and Engineering*, vol. 6, no. 1, pp. 16–24, 2009.

[MAR 03] MARROCCO G., MATTIONI L., CALABRESE C., "Multiport sensor RFIDs for wireless passive sensing of objects – basic theory and early results", *IEEE Transactions on Antennas and Propagation*, vol. 51, no. 1, pp. 31–39, 2003.

[MAS 13] MASHAAL O.A., RAHIM S.K.A., ABDULRAHMAN A.Y. *et al.*, "A coplanar waveguide fed two arm Archimedean spiral slot antenna with improved bandwidth", *IEEE Transactions on Antennas and Propagation*, vol. 61, no. 2, pp. 939–943, 2013.

[MAT 98] MATZLER C., "Microwave permittivity of dry sand", *IEEE Transactions on Geoscience and Remote Sensing*, vol. 36, no. 1, pp. 317–319, 1998.

[MAZ 11] MAZAR H., "A comparison between European and North American wireless regulations", *Technical Symposium at ITU Telecom World*, Geneva, pp. 182–186, 2011.

[MCV 06] MCVAY J., HOORFAR A., ENGHETA N., "Space-filling curve RFID tags", *IEEE Radio and Wireless Symposium Digest*, pp. 199–202, 2006.

[MEY 06] MEYER-BASE U., NATARAJAN H., CASTILLO E. *et al.*, "Faster than the FFT: the chirp-z RAG-n discrete fast Fourier transform", *Frequenz*, vol. 60, nos. 7–8, pp. 147–151, 2006.

[MIC 14] MICROCHIP, "PIC 16(L)F1827 8-bit flash MCU with nanowatt XLP", available at http://www.microchip.com/wwwproducts/Devices.aspx?dDocName=en538963, 2014.

[MIC 14] MICROWAVE VISION GROUP, Open Boundary Quad Ridge Horns, Datasheet (QH2000), available at http://www.mvg-world.com/system/files/Open%20Boundary%20Quad-Ridge_bd.pdf, 2014.

[MOL 03] MOLESA S., REDINGER D.R., HUANG D.C. *et al.*, "High-quality ink-jet-printed multilevel interconnects and inductive components on plastic for ultra-low-cost RFID applications", *Proceedings of MRS*, vol. 769, p. H8.3, 2003.

[MOR 93] MORI Y., ELLINGWOOD B.R., "Reliability-based service-life assessment of aging concrete structures", *Journal of Structural Engineering*, vol. 119, no. 5, pp. 1600–1621, 1993.

[MUK 07] MUKHERJEE S., "Chipless radio frequency identification by remote measurement of complex impedance", *Proceedings of 37th European Microwave Conference*, Munich, Germany, pp. 1007–1010, 2007.

[NAP 11] NAPHADE M., BANAVAR G., HARRISON C. *et al.*, "Smarter cities and their innovation challenges", *Computer*, vol. 44, no. 6, pp. 32–39, 2011.

[NIK 07] NIKITIN P.V., RAO K.V.S., MARTINEZ R.D., "Differential RCS of RFID Tag", *Electronics Letters*, vol. 43, no. 8, pp. 431–432, 2007.

[NOV 12] NOVELDAAS, "NVA 6100", available at http://novelda.no/content/nva6100, 2012.

[NXP 04a] NXP SEMICONDUCTORS, "BAP51-03 general purpose PIN diode", available at http://www.nxp.com/documents/data_sheet/BAP51-03.pdf, 2004.

[NXP 04b] NXP SEMICONDUCTORS, "BAP64-03 general purpose PIN diode", available at http://www.nxp.com/documents/data_sheet/BAP64-03.pdf, 2004.

[OCC 11] OCCHIUZZI C., RIDA A., MARROCCO G. *et al.*, "RFID passive gas sensor integrating carbon nanotubes", *IEEE Transactions on Microwave Theory and Techniques*, vol. 59, no. 10, pp. 2674–2684, 2011.

[OHB 04] OHBUCHI E., HANAIZUMI H., HOCK L.A., "Barcode readers using the camera device in mobile phones", *International Conference on Cyberworlds*, Washington, USA, pp. 260–265, November 2004.

[OLS 13] OLSZEWSKI B., FENTON S., TWOREK B. *et al.*, "RFID positioning robot: an indoor navigation system", *IEEE International Conference on Electro/Information* Technology, May 2013.

[ONG 02] ONG K.G., ZENG K., GRIMES C.A., "A wireless, passive carbon nanotube-based gas sensor", *IEEE Sensors Journal*, vol. 2, no. 2, pp. 82–88, 2002.

[ONG 08] ONG J.B., YOU Z., MILLS-BEALE J. *et al.*, "A wireless, passive embedded sensor for real-time monitoring of water content in civil engineering materials", *IEEE Sensors Journal*, vol. 8, no. 12, pp. 2053–2058, 2008.

[OPA 06] OPASJUMRUSKIT K., THANTHIPWAN T., SATHUSEN O. *et al.*, "Self-powered wireless temperature sensors exploit RFID technology", *IEEE Pervasive Computing*, vol. 5, no. 1, pp. 54–61, 2006.

[PAL 07] PALMER R.C., *The Bar Code Book: Fifth Edition – A Comprehensive Guide To Reading, Printing, Specifying, Evaluating, And Using Bar Code and Other Machine-Readable Symbols*, Trafford Publishing, 2007.

[PAR 05] PARADISO J.A., STARNER T., "Energy scavenging for mobile and wireless electronics", *IEEE Pervasive Computing*, vol. 4, no. 11, pp. 18–27, 2005.

[PAR 13] PARK S., LEE H., "Self-recognition of vehicle position using UHF passive RFID tags", *IEEE Transactions on Industrial Electronics*, vol. 60, no. 1, pp. 226–234, 2013.

[PET 12] PETROFF A., "A pratical, high performance ultra-wideband radar platform", *IEEE Radar Conference*, pp. 880–884, 2012.

[POU 03] POURVOYEUR K., STELZER A., OSSBERGER G. *et al.*, "Wavelet-based impulse reconstruction in UWB-radar", *IEEE MTT-S International Microwave Symposium*, pp. 603–606, 2003.

[POU 11] POURAHMADAZAR J., GHOBADI C., NOURINIA J. *et al.*, "Broadband CPW-fed circularly polarized square slot antenna with inverted-L strips for UWB applications", *IEEE Antennas and Wireless Propagation Letters*, vol. 10, pp. 369–372, 2011.

[PRE 07] PRERADOVIC S., BALBIN I., KARMAKAR N.C., "Development of a low-cost semi-passive transponder for sensor applications at 2.4 GHz", *International Symposium on Communications and Information Technologies*, pp. 131–135, October 2007.

[PRE 09] PRERADOVIC S., BALBIN I., KARMAKAR N.C. *et al.*, "Multiresonator-based chipless RFID system for low-cost item tracking", *IEEE Transactions on Microwave Theory and Techniques*, vol. 57, no. 5, pp. 1411–1419, 2009.

[PUR 08] PURSULA P., VAHA-HEIKKILA T., MULLER A. *et al.*, "Millimeter-wave identification – a new short-range radio system for low-power high data-rate applications", *IEEE Transactions on Microwave Theory and Techniques*, vol. 56, no. 10, pp. 2221–2228, 2008.

[RAB 04] RABBACHIN A., OPPERMANN I., "Synchronization analysis for UWB systems with a low-complexity energy collection receiver", *Proceedings of IEEE Ultra wide band Systems and Technologies Conference*, pp. 288–292, 2004.

[RAM 11] RAMOS A., LAZARO A., GIRBAU D. *et al.*, "Time-domain measurement of time-coded UWB chipless RFID tags", *Progress in Electromagnetics Research*, vol. 116, pp. 313–331, 2011.

[RAO 05] RAO K.V.S., NIKITIN P.V., LAM S.M., "Antenna design for UHF RFID tags: a review and a practical application", *IEEE Transactions on Antennas Propagation*, vol. 53, no. 12, pp. 3870–3876, 2005.

[REI 01] REINDL L., RUPPEL C.C.W., BEREK S. *et al.*, "Design, fabrication, and application of precise SAW delay lines used in an FMCW radar system", *IEEE Transactions on Microwave Theory and Techniques*, vol. 49, no. 4, pp. 787–794, 2001.

[REI 04] REINDL L.M., SHRENA I.M., "Wireless measurement of temperature using surface acoustic waves sensors", *IEEE Transactions on Ultrasonics, Ferroelectrics and Frequency Control*, vol. 51, no. 11, pp. 1457–1463, 2004.

[REI 98] REINDL L., SCHOLL G., OSTERTAG T. *et al.*, "Theory and application of passive SAW radio transponders as sensors", *IEEE Transactions on Ultrasonics, Ferroelectrics, and Frequency Control*, vol. 45, no. 5, pp. 1281–1292, 1998.

[RHI 98] RHIM H.C., BUYUKOZTURK O., "Electromagnetic properties of concrete at microwave frequency range", *ACI Materials Journal*, vol. 95, no. 3, pp. 262–271, 1998.

[RIA 09] RIDA A., YANG L., VYAS R., *et al.*, "Conductive inkjet-printed antennas on flexible low-cost paper-based substrates for RFID and WSN applications", *IEEE Antennas and Propagation Magazine*, vol. 51, no. 3, pp. 13–29, 2009.

[SAM 08] SAMPLE A.P., YEAGER D.J., POWLEDGE S.P. *et al.*, "Design of an RFID-based battery-free programmable sensing platform", *IEEE Transactions on Instrumentation and Measurement*, vol. 57, no. 11, pp. 2608–2615, 2008.

[SCH 09] SCHULER M., MANDEL C., MAASCH M. *et al.*, "Phase modulation scheme for chipless RFID and wireless sensor tags", *Proceedings of Asia Pacific Microwave Conference*, pp. 229–232, 2009.

[SHA 11] SHAKER G., SAFAVI-NAEINI S., SANGARY N. *et al.*, "Inkjet printing of ultrawideband (UWB) antennas on paper-based substrates", *IEEE Antennas and Wireless Propagation Letters*, vol. 10, pp. 111–114, 2011.

[SHR 07] SHRETHA S., VEMAGIRI J., AGARWAL M. *et al.*, "Transmission line delay-based radio frequency identification (RFID) tag", *Microwave and Optical Technology Letters*, vol. 49, no. 8, pp. 1900–1904, 2007.

[SHR 09] SHRESTHA S., BALACHANDRAN M., AGARWAL M. *et al.*, "A chipless RFID sensor system for cyber centric monitoring applications", *IEEE Transactions on Microwave Theory and Techniques*, vol. 57, no. 5, pp. 1303–1309, 2009.

[SHU 01] SUH Y.H., PARK I., "A broadband eccentric annular slot antenna", *IEEE Antennas and Propagation Society International Symposium*, vol. 1, pp. 94–97, 2001.

[SIV 08] SIVARAMAKRISHNAN S., RAJAMANI R., SMITH C.S. *et al.*, "Carbon nanotube-coated surface acoustic wave sensor for carbon dioxide sensing", *Sensors and Actuators B*, vol. 132, pp. 296–304, 2008.

[SKY 07] SKYWORKS INC., GaAs IC high-isolation positive control SPDT switch non-reflective switch LF-4GHz, Data sheet, October 2007.

[SMI 11] SMITH P., *Comparing Low-Power Wireless Technologies,* Digikey Electronics Article Library, available at http://www.digikey.com/en/articles/techzone/2011/avg/comparing-low-power-wireless-technologies, California, USA, 2011.

[SUW 14] SUWALAK R., PHONGCHAROENPANICH C., TORRUNGRUENG D. *et al.*, "Determination of dielectric property of construction material products using a novel RFID sensor", *Progress in Electromagnetics Research*, vol. 130, pp. 601–617, 2014.

[SVA 92] SVACINA J., "A simple quasi-static determination of basic parameters of multilayer microstrip and coplanar waveguide", *IEEE Microwave and Guided Wave Letters*, vol. 2, no. 10, pp. 385–387, 1992.

[SZE 10] SZE J-Y., HSU C-I.G., CHEN Z-W. *et al.*, "Broadband CPW-fed circularly polarized square slot antenna with lightening-shaped feedline and inverted-L grounded strips", *IEEE Transactions on Antennas and Propagation*, vol. 58, no. 3, pp. 973–977, 2010.

[TAN 94] TANAKA M., SUZUKI R., SUZUKI Y. *et al.*, "Microstrip antenna with solar cells for microsatellites", *Proceedings of the IEEE International Symposium on Antennas and Propagation (AP-S)*, vol. 2, pp. 786–789, June 1994.

[TAV 07] TAVASSOLIAN N., NIKOLAOU S., TENTZERIS M.M., "A flexible UWB elliptical slot antenna with a tuning uneven U-shape stub on LCP for microwave tumor detection", *Asia-Pacific Microwave Conference*, pp. 1–4, December 2007.

[TAY 12] TAYLOR J.D., *Ultrawideband Radar: Applications and Design*, CRC Press, USA, 2012.

[TED 13] TEDJINI S., KARMAKAR N.C., PERRET E. *et al.*, "Hold the chips: chipless technology, an alternative technique for RFID", *IEEE Microwave Magazine*, vol. 14, no. 5, pp. 56–65, 2013.

[TES 10] TESORIERO R., TEBAR R., GALLUD J.A. *et al.*, "Improving location awareness in indoor spaces using RFID technology", *Expert Systems with Applications*, vol. 37, no. 1, pp. 894–898, 2010.

[TEX 12] TEXAS INSTRUMENTS, MSP430FR59XX mixed-signal microcontrollers, report SLAS704D, October 2012.

[TEX 14] TEXAS INSTRUMENTS, "Wireless connectivity", available at http://www.ti.com/lit/sg/slab056d/slab056d.pdf, 2014.

[THA 10] THAI T.T., MEHDI J.M., AUBERT H. *et al.*, "A novel passive wireless ultrasensitive RF temperature transducer for remote sensing", *IEEE MTT-S Microwave Symposium Digest*, California, USA, pp. 473–476, May 2010.

[THA 11] THAI T.T., YANG L., DEJEAN G.R. *et al.*, "Nanotechnology enables wireless gas sensing", *IEEE Microwave Magazine*, pp. 84–95, 2011.

[TIM 14] TIME DOMAIN, "PulsON P400 MRM", available at http://www.timedomain.com/p400-mrm.php, 2014.

[VEN 11] VENA A., PERRET E., TEDJINI S., "Chipless RFID tag using hybrid coding technique", *IEEE Transactions on Microwave Theory and Techniques*, vol. 59, no. 12, pp. 3356–3364, 2011.

[VEN 12] VENA A., PERRET E., TEDJINI S. *et al.*, "A compact chipless RFID tag with environment sensing capability", *2012 IEEE MTT-S International Microwave Symposium Digest*, Montreal QC, Canada, pp. 1–3, June 2012.

[VEN 13a] VENA A., PERRET E., TEDJINI S., "A depolarizing chipless RFID tag for robust detection and its FCC compliant UWB reading system", *IEEE Transactions on Microwave Theory and Techniques*, vol. 61, no. 8, pp. 2982–2994, 2013.

[VEN 13b] VENA A., SYDÄNHEIMO L., TENTZERIS M.M. *et al.*, "A novel inkjet printed carbon nanotube-based chipless RFID sensor for gas detection", *IEEE Proceedings of the 43rd European Microwave Conference*, Nuremberg, Germany, October 2013.

[VII 09] VIIKARI V., SEPPA H., "RFID MEMS sensor concept based on intermodulation distortion", *IEEE Sensors Journal*, vol. 9, no. 12, pp. 1918–1923, 2009.

[VIR 11] VIRTANEN J., UKKONEN L., BJÖRNINEN T. *et al.*, "Temperature sensor tag for passive UHF RFID systems", *IEEE Sensors and Applications Symposium*, San Antonio, TX, USA, pp. 312–317, February 2011.

[VIS 14] VISHAY B., Platinum SMD Flat Chip Temperature Sensor, available at http://www.vishay.com/docs/28762/28762.pdf, 2014.

[VIT 05] VITA G.D., IANNACCONE G., "Design criteria for the RF section of UHF and microwave passive RFID transponders", *IEEE Transactions on Microwave Theory and Techniques*, vol. 53, no. 9, pp. 2978–2990, 2005.

[VIT 10] VITAZ J.A., BUERKLE A.M., SARABANDI K., "Tracking of metallic objects using a retro-reflective array at 26 GHz", *IEEE Transactions on Antennas and Propagation*, vol. 58, no. 11, pp. 3539–3544, 2010.

[VYK 09] VYKAS R., LAKAFOSIS V., RIDA A. *et al.*, "Paper-based RFID-enabled wireless platforms for sensing applications", *IEEE Transactions on Microwave Theory and Techniques*, vol. 57, no. 5, pp. 1370–1382, 2009.

[WAN 04] WANT R., "Enabling ubiquitous sensing with RFID", *Computer*, vol. 37, no. 4, pp. 84–86, 2004.

[WAN 06] WANT R., "An introduction to RFID technology", *IEEE Pervasive Computing*, vol. 5, no. 1, p. 25–33, 2006.

[WAN 13] WANT R., SCHILIT B., LASKOWSKI D., "Bluetooth LE finds its niche", *IEEE Pervasive Computing*, vol. 12, no. 4, pp. 12–16, 2013.

[WEH 10] WEHRLI S., GIERLICH R, HÜTTNER J. *et al.*, "Integrated active pulsed reflector for an indoor local positioning system", *IEEE Transactions on Microwave Theory and Techniques*, vol. 58, no. 2, Valletta, Malta, pp. 267–276, April 2010.

[WEL 09] WELBOURNE E., BATTLE L., COLE G. *et al.*, "Building the Internet of things using RFID: the RFID ecosystem experience", *IEEE Internet Computing*, vol. 13, no. 13, pp. 48–55, 2009.

[WHE 07] WHEELER A., "Commercial applications of wireless sensor networks using Zigbee", *IEEE Communications Magazine*, vol. 45, no. 4, pp. 70–77, 2007.

[WON 10] WONG S.M., TAN CHONG E. *et al.*, "Efficient RFID tag placement framework for in building navigation system for the blind", *8th Asia-Pacific Symposium on Information and Telecommunication Technologies*, June 2010.

[XET 15] XETHRU BY NOVELDA, "XeThru X2 and XeThru X1", available at https://www.xethru.com/content/technology-0, 2015.

[XU 07] XU R.P., HUANG X.D., CHENG C.H., "Broadband circularly polarized wide-slot antenna", *Microwave and Optical Technology Letters*, vol. 49, no. 5, pp. 1005–1007, 2007.

[YAN 08] YANG L., MARTIN L.J., STAICULESCU D. *et al.*, "Conformal magnetic composite RFID for wearable RF and bio-monitoring applications", *IEEE Transactions on Microwave Theory and Techniques*, vol. 56, no. 12, pp. 3223–3230, 2008.

[YAN 09] YANG L., ZHANG R., STAICULESCU R. *et al.*, "A novel conformal RFID-enabled module utilizing inkjet-printed antennas and carbon nanotubes for gas-detection applications", *IEEE Antennas and Wireless Propagation Letters*, vol. 8, pp. 653–656, 2009.

[YIC 08] YICK Y., MUKHERJEE B., GHOSAL D., "Wireless sensor network survey", *Computer Networks*, vol. 52, pp. 2292–2330, 2008.

[YIN 10] YIN J., YI J., LAW M.K. *et al.* "A system-on-chip EPC Gen-2 passive UHF RFID tag with embedded temperature sensor", *IEEE Journal of Solid-State Circuits*, vol. 45, no. 11, pp. 2404–2420, 2010.

[ZAN 11] ZANOLLI Z., LEGHRIB R., FELTEN A. *et al.*, "Gas sensing with Au-decorated carbon nanotubes", *ACS Nano*, vol. 6, pp. 4529–4599, 2011.

[ZHA 06] ZHANG L., RODRIGUEZ S., TENHUNEN H. *et al.*, "An innovative fully printable RFID technology based on high speed time-domain reflection", *Conference on High Density Microsystem Design and Packaging and Component Failure Analysis*, pp. 166–170, 2006.

[ZHE 14] ZHENG Y.C., HUA G., ZHENG Y.L. *et al.*, "Design of a miniaturized RFID tag antenna with BAP technique", *3rd Asia-Pacific Conference on Antennas and Propagation*, pp. 256–258, April 2014.

[ZIT 10] ZITO F., FRAGOMENI L., AQUILINO F. *et al.*, "Wireless temperature sensor integrated circuits with on-chip antennas", *15th IEEE Mediterranean Electrotechnical Conference*, Valletta, Malta, pp. 1368–1373, 2010.

[ZOU 11] ZOU Z., SARMIENTO D., WANG P. *et al.*, "A low-power and flexible energy detection IR-UWB receiver for RFID and wireless sensor networks", *IEEE Transactions on Circuits and Systems*, vol. 58, no. 7, pp. 1470–1482, 2011.

Index

Printed in the United States
By Bookmasters